Electromagnetism and Interconnections

Electromagnetism and Interconnections

Advanced Mathematical Tools for Computer-aided Simulation

Stéphane Charruau

Series Editor
Pierre-Noël Favennec

First published in Great Britain and the United States in 2009 by ISTE Ltd and John Wiley & Sons, Inc.

Apart from any fair dealing for the purposes of research or private study, or criticism or review, as permitted under the Copyright, Designs and Patents Act 1988, this publication may only be reproduced, stored or transmitted, in any form or by any means, with the prior permission in writing of the publishers, or in the case of reprographic reproduction in accordance with the terms and licenses issued by the CLA. Enquiries concerning reproduction outside these terms should be sent to the publishers at the undermentioned address:

ISTE Ltd
27-37 St George's Road
London SW19 4EU
UK

John Wiley & Sons, Inc.
111 River Street
Hoboken, NJ 07030
USA

www.iste.co.uk

www.wiley.com

© ISTE Ltd, 2009

The rights of Stéphane Charruau to be identified as the author of this work have been asserted by him in accordance with the Copyright, Designs and Patents Act 1988.

Library of Congress Cataloging-in-Publication Data

Charruau, Stéphane.
  Electromagnetism and interconnections : advanced mathematical tools for computer-aided simulation / Stéphane Charruau.
      p. cm.
  Includes bibliographical references and index.
  ISBN 978-1-84821-107-0
  1. Telecommunication lines--Computer simulation. 2. Electromagnetic waves--Mathematical models. I. Title.
  TK5103.15.C45 2009
  621.381--dc22
                                                                                        2008043382

British Library Cataloguing-in-Publication Data
A CIP record for this book is available from the British Library
ISBN: 978-1-84821-107-0

Printed and bound in Great Britain by CPI Antony Rowe, Chippenham, Wiltshire.

Table of contents

Acknowledgements xi

Introduction. xiii

Chapter 1. Theoretical Foundations of Electromagnetism. 1

1.1. Elements of the theory of distributions applied to
electromagnetism 1
   1.1.1. Choosing a presentation of the foundations of
   electromagnetism 1
   1.1.2. Linear modeling of physical laws and Green's kernels 2
   1.1.3. Accounting for the "natural symmetries" of physical laws. 3
   1.1.4. Motivation for using the theory of distributions 4
   1.1.5. Quick review of the theory of distributions. 5
   1.1.6. Application to electromagnetism 9
1.2. Vector analysis review according to the theory of distributions 11
   1.2.1. Derivation of discontinuous functions defined on R 11
   1.2.2. Derivative of linear mappings. 12
   1.2.3. Derivation of discontinuous functions on a surface in $\Re^3$ 12
   1.2.4. Derivation of vector distributions in $\Re^3$. 13
   1.2.5. Algebra of the operator ∇ 13
1.3. Maxwell's equations according to the theory of distributions 14
   1.3.1. Symmetries and duality in electromagnetism 14
   1.3.2. The symmetry laws of distributions in electromagnetism 14
   1.3.3. Application to the first couple of Maxwell's equations. 15
   1.3.4. Behavior law of materials by means of the theory of
   distributions 19
   1.3.5. Application to the second couple of Maxwell's equations 19
   1.3.6. Charge density, current density, continuity equations. 20
   1.3.7. Integral form of Maxwell's equations 22
1.4. Conclusion 24

Chapter 2. Full Wave Analysis 25

2.1. Discontinuities in electromagnetism 25
    2.1.1. Initial and boundary conditions according to the theory
    of distributions 25
    2.1.2. Electromagnetic images, incident and reflected fields 28
    2.1.3. Method of moments for the numerical computation of
    electromagnetic fields 29
2.2. Potentials in electromagnetism 33
    2.2.1. Scalar and vector potentials, duality between electrical and
    magnetic potentials. 33
    2.2.2. Lossy propagation equations, the Lorentz gauge 35
    2.2.3. Green's kernels for harmonic electromagnetic waves in
    heterogenous media 39
2.3. Topology of electromagnetic interferences 42
    2.3.1. Introduction 42
    2.3.2. Topological modeling of electromagnetic interferences 43
    2.3.3. Partitioning the electrical network in respect of electromagnetic
    interferences 45
    2.3.4. The tree of electromagnetic interferences and the
    problem of loops 46
2.4. Conclusion 50

Chapter 3. Electromagnetism in Stratified Media 51

3.1. Electrical and magnetic currents in stratified media 52
    3.1.1. Scope of the theory, defining stratified media 52
    3.1.2. Integral formulation of the current derivative versus time:
    general case 53
    3.1.3. Integral formula of the current derivative relative to space in the
    direction of the vector potential 61
    3.1.4. Duality between electrical and magnetic currents in
    lossless media. 63
3.2. Straight stratified media 67
    3.2.1. Scope 67
    3.2.2. Lossy propagation equations and the variational approach. . . . 67
    3.2.3. Spectral analysis of the longitudinal field. 71
    3.2.4. From Maxwell's equations to transmission line equations 76
    3.2.5. Generalized transmission line matrix equation. 79
    3.2.6. Non-existence of the TM and TE modes separately. 81
    3.2.7. Electrical (or magnetic) currents 84
3.3. Conclusion 84

Chapter 4. Transmission Line Equations 85

4.1. Straight homogenous dielectric media with lossless conductors. 86
    4.1.1. Hypothesis 86
    4.1.2. Electrical current formulae in TM mode
    of propagation. 86
    4.1.3. Magnetic current formulae in TE mode of propagation 89
    4.1.4. Spectral analysis of electromagnetic fields 89
    4.1.5. Modal analysis of electrical current and lineic charge 96
    4.1.6. Modal analysis of scalar and vector potentials 101
    4.1.7. Transmission line with distributed sources corresponding to a
    waveguide 103
4.2. TEM mode of wave propagation 104
    4.2.1. Defining the TEM mode and the transmission lines. 104
    4.2.2. Basic existence condition of a TEM propagation mode 105
    4.2.3. Variational numerical computation of the lowest wavelength. . . 107
    4.2.4. Telegrapher's equation for current and electrical charge
    per unit length. 109
    4.2.5. Lorentz condition and telegrapher's equation for vector
    potentials and scalars in TEM mode. 111
    4.2.6. Lineic distribution of electrical charges and the
    Poisson equation 112
    4.2.7. Transmission line equations for lossy dielectrics and
    lossless conductors. 115
    4.2.8. Green's kernels and the numerical computation of
    lineic parameters 117
4.3. Quasi-TEM approximation for lossy conductors and dielectrics. 122
    4.3.1. Foucault's modal currents of electromagnetic field
    propagation in lossy media 122
    4.3.2. Quasi-TEM approximation of coupled lossy
    transmission lines. 124
4.4. Weakly bent transmission lines in the quasi-TEM approximation. . . . 126
    4.4.1. Bent lossy heterogenous media with lossless conductors. 126
    4.4.2. Bent lossy homogenous media with lossless conductors 127
    4.4.3. Bent lossless conductors such that e_n does not depend on
    q_l, and e_l and C_H do not depend on q_n. 128
    4.4.4. Lineic capacitance tied to a weak curvature of
    a transmission line 128
4.5. Conclusion 130

Chapter 5. Direct Time-domain Methods 131

5.1. "Direct" methods in the time domain. 132
    5.1.1. Defining a "direct" method in the time domain 132

5.1.2. Single lossless transmission lines in homogenous media. 132
5.2. Lossless coupled transmission lines in homogenous media 143
    5.2.1. Homogenous coupling 143
    5.2.2. Heterogenous coupling. 150
    5.2.3. Bifurcations 151
    5.2.4. Complex distributed parameter networks 156
    5.2.5. Estimation of the transient state time of signals 159
    5.2.6. Numerical computation of the characteristic
    impedance matrix. 161
5.3. Conclusion 162

Chapter 6. Discretization in the Time Domain 163

6.1. Finite difference method in the time domain 163
    6.1.1. From full wave analysis to nodal operational matrices 163
    6.1.2. Recursive differential transmission line matrix equation
    of complex networks. 167
    6.1.3. Estimation of the transient state time 168
    6.1.4. Finite difference approximation of differential
    operators in the time domain 170
    6.1.5. Application to lumped quadripole modeling approximation
    in the time domain 173
    6.1.6. Complex distributed and lumped parameter
    networks approximation. 175
6.2. Matrix velocity operator interpolation method 179
    6.2.1. Difficulties set by the compounded matrix functions 179
    6.2.2. Matrix velocity matrix operator of stratified
    heterogenous media 181
    6.2.3. Matrix velocity operator interpolation method for
    the matrix drift equation 183
6.3. Conclusion 187

Chapter 7. Frequency Methods. 189

7.1. Laplace transform method for lossy transmission lines 190
    7.1.1. Transfer matrix in the Laplace domain 190
    7.1.2. Transfer impedance matrix, impedance matching,
    scattering matrix 198
7.2. Coming back in the time domain 202
    7.2.1. Inverse Laplace transform for lossy transmission lines. 202
    7.2.2. Method of the contribution of loops 203
    7.2.3. Application to the distortion of a Dirac pulse in lossy media . . . 206
    7.2.4. Classical kernel of the convolution methods 207
    7.2.5. Diffusion equation and the time-varying "skin depth" 208

7.2.6. Multiple reflections processing 209
7.3. Method of the discrete Fourier transform 210
   7.3.1. Fourier transform and the harmonic steady state. 210
   7.3.2. Discrete Fourier transform and the sampling procedure 211
   7.3.3. Application to digital signal processing 213
   7.3.4. Bifurcations and complex networks of lossy
   transmission lines 215
7.4. Conclusion 217

Chapter 8. Time-domain Wavelets 219

8.1. Theoretical introduction 219
   8.1.1. Motivation for the time-domain wavelets method. 219
   8.1.2. General mathematical framework 220
   8.1.3. Seed and generator of direct and reverse wavelets family 221
8.2. Application to digital signal propagation 226
   8.2.1. Application to lossless guided wave analysis in
   the time domain. 226
   8.2.2. Application to the telegrapher's equation 230
   8.2.3. Convergence of wavelet expansions, numerical
   approximation. 233
8.3. Conclusion 241

Chapter 9. Applications of the Wavelet Method 243

9.1. Coupled lossy transmission lines in the TEM approximation 243
   9.1.1. Wavelets in homogenously coupled lossy transmission lines . . . 243
   9.1.2. Multiple reflections into lossy coupled lines 250
   9.1.3. Comparative analysis of frequency and wavelets methods. 255
9.2. Extension to 3D wavelets and electromagnetic perturbations 256
   9.2.1. Basic second-order partial differential equation
   of electromagnetic waves 256
   9.2.2. Obtaining the wavelet generating equation: $Au = \Lambda u$. 257
   9.2.3. Direct and reverse generators of the wavelet base. 258
   9.2.4. Spherical seed and wavelets having a zero divergence 260
   9.2.5. Modeling electromagnetic perturbations in lossy media 261
   9.2.6. Guided propagation in interconnection structures 262
9.3. Conclusion 262

Appendices 263

   Appendix A. Physical Data 263
   Appendix B. Technological Data. 267
   Appendix C. Lineic Capacitors 269

Appendix D. Modified Relaxation Method. 275
Appendix E. Cylindrical Wavelets. 277
Appendix F. Wavelets and Elliptic Operators 281

References 287

Index 291

Acknowledgments

Before anything else, the author would like to dedicate this book to his wife Chantal and his daughters Coralie, Laure, Olivia and Héloïse to whom the author pays homage for their emotional support during the many years of work on this book.

The author expresses his appreciation to Professor André Touboul of the University of Bordeaux and researcher at the Laboratoire d'Etudes de l'Intégration des Composants et Systèmes Electroniques for scientific discussions regarding mathematical methods and their application to interconnections modeling.

The author thanks Mrs. Hélène Misson PhD of the University of Bordeaux for help in computations, curve generation, text typing and mathematical equation editing.

The author extends thanks to Mr. Olivier Meili, graduate student of the University of Bordeaux, for having produced illustrations and carrying out a typing review.

The author acknowledges Mr. Dominique Gili-Lacoste from the society Trièdre Concept for help in three-dimensional image generation.

Introduction

This book is intended for scientists, research engineers and graduate students interested in electromagnetism, microwave theory, electrical engineering and the development of simulation tool software devoted to very high-speed electronic system design automation or in the application of mathematics to these topics.

The subject matter of this book concerns the theoretical problems of modeling the electrical behavior of the interconnections met in electronic products that have become ubiquitous in daily life since the end of the 20th century. Most electronic products have digital processors that have more and more inner and outer conductors with smaller and smaller geometries. This means increasingly more parasitic electromagnetic effects occur inside and outside these processors that then cannot work correctly. The aim of this book is to show the theoretical tools of waveform prediction at the design stage of a complex and high-speed digital electronic system.

The topics that the book covers are not new; indeed transmission line analysis has a very long history. In the middle of the 19th century, the technical problems posed by the bad quality of transmissions traveling through submarine cables led telecommunications engineers, mainly those working in Britain, to become interested in Maxwell's theory published in 1873. One of their number, Heaviside, formulated Maxwell's equations in their modern form in 1884. Since that time and up to Word War II, Heaviside's formalism greatly supported the development of wireless transmissions, then that of radars and their waveguide technology, while Lord Kelvin's line modeling based on chaining a huge number of lumped electrical networks was used in phone transmissions analysis.

The first theoretical approach of two weakly coupled lines based on Kelvin's line modeling was presented by Dr Campbell in 1912. The extension of this work to the most general case of strongly coupled line equations using matrix functions came in 1937 with the work of L. A. Pipes. Then in 1947, M. Cotte studied the propagation of pulses along two coupled lossless lines by means of Heaviside's operational calculus in order to understand the results of electrical perturbation measurements of power mains.

In 1955, the "stripline" and "microstrip" techniques established a bridge between the world of the classical transmission lines and that of waveguides. At that time, S. A. Schelkunoff proposed a theory of the TEM (transverse electrical magnetic) mode of lossless propagation based on a conversion of Maxwell's equations into lossless transmission line equations, thus completing the previously mentioned bridge. L. Brillouin studied rigorous theoretical approaches to the losses inside transmission lines from 1932 to 1960 but the solution remained difficult to apply by researchers involved in the development of digital electronics since 1963.

Nowadays, the electrical behavior of lossy lines is modeled in accordance with the assumption of the so-called "quasi-TEM" approximation by means of the modal analysis of the transmission line matrix equation in the frequency domain. The FFT (fast Fourier transform) method is widely used now in the software packages within CAD (computer-aided design) systems devoted to the industrial development of electronic modules. This latter method cannot handle the nonlinearities set by the electrical behavior of semiconductors used by these modules. Handling nonlinearities in lossy lines requires classically the time domain *convolution* method that uses too much computer time and memory space. Furthermore, modern substrates needed by these electronic modules can be flexible and bent. In any case, the network of interconnections lies on stacked layers linked by vertical conductors called "vias". The curvature of modern substrates and the vias leads to the need for three-dimensional (or 3D) modeling.

The book is divided into nine chapters, each one beginning with an introductory passage giving the leading thread of the chapter. A brief conclusion aimed at the most important results is presented at the end of each chapter. A glossary of terms and a list of references appear at the end of the book.

Chapter 1 is devoted to the theoretical foundations of electromagnetism in terms of highlighting the natural symmetries between the distributions met in electromagnetism.

Chapter 2 concerns full wave analysis based on Maxwell's equations and their boundary conditions with an original topological approach to electromagnetic interferences.

Chapter 3 is devoted to the equations of electromagnetism in "stratified media", even those being bent, where up-to-date electronic interconnections are designed.

Chapter 4 is devoted to the transformation of these equations into transmission line equations, including an original skin effect modeling suited to the design of interconnections.

Chapter 5 concerns the direct time domain methods compatible with handling nonlinearity in complex lossless networks, using advanced powerful matrix methods.

Chapter 6 concerns the discretization process in the time domain needed for the cases of lumped circuits between transmission lines or in heterogenous media.

Chapter 7 deals with the frequency methods which account naturally for the losses in dielectrics and conductors as well as in complex networks, even with bifurcations.

Chapter 8 presents the new time-domain wavelets which are well suited for high-speed digital signals running in complex lossy nonlinear networks.

Chapter 9 presents the applications of the new time-domain wavelets to lossy coupled lines and the problems of 3D electromagnetic perturbations.

Chapter 1

Theoretical Foundations of Electromagnetism

This first chapter is devoted to a brief overview of electromagnetism needed for modeling the electrical behavior of interconnections traveled by very high speed digital or pulsed signals. After a quick look at the historical development of electromagnetism, the tremendous interest in the theory of distributions applied to electromagnetism is highlighted in terms of digital signal transmission analysis. Then, the strictly useful elements of this theory and the necessary vector analysis are discussed, thus allowing an original derivation of Maxwell's equations from the intuitive geometric properties of the linear relations between electromagnetic features which are modeled by polar and axial vector distributions. The integral forms of Maxwell's equations are presented.

1.1. Elements of the theory of distributions applied to electromagnetism

1.1.1. *Choosing a presentation of the foundations of electromagnetism*

Electromagnetism phenomena [ROC] [JAC1], the foundations of transmission lines analysis, concern the interactions between electricity and magnetism in nature observed experimentally by Oersted in 1819 (magnetic field created by an electrical current) and by Faraday in 1830 (electrical current created by a variable magnetic field), completed by the propagation of electromagnetic waves discovered by Hertz in 1887. The results of Oersted's experiments were translated into mathematical laws by Biot and Savart in 1820 and then by Ampere, and those of Faraday by Lenz and then by Foucault in 1850, thus leading to Maxwell's theory of 1873, the equations of which got their final form thanks to Heaviside in 1884.

Electromagnetism was developed from the experimental results obtained during the first part of the 19th century, continuing to Maxwell's equations and Hertz waves at the end of the century, so most of the classical presentations follow the historical approach.

Nowadays, the dramatic development of telecommunications, radiodetection technology and media electronics has given rise to a whole set of new experiments that confirm Maxwell's theory.

Yet, the present challenge is to develop the technology using electromagnetic features at the lowest cost. This has to be achieved by means of reliable computational prediction of performance. The complexity of technological structures requires the best algorithms suited for solving Maxwell's equations or any equivalent form.

Thus, our presentation is aimed at the mathematical tools of the modeling of electromagnetic phenomena with respect to the technological structures related to interconnections inside electronic boards.

1.1.2. *Linear modeling of physical laws and Green's kernels*

Electromagnetic phenomena are mainly linear, which means the effects are proportional to the causes and the sum of causes gives the sum of effects. Therefore, the results of measurements of several electromagnetic features show "linear relations" between the features that are valid inside the boundary of the domain where the phenomena involved are studied. This is the reason why these features can be depicted efficiently as elements of a vector space which are defined by the availability of the following operations: internal addition and external multiplication by a real number.

Moreover, the total energy involved in these phenomena is finite which leads to their physical features being represented mathematically by square integrable functions. According to the concept used in mathematics, it is said these functions have to belong to the so-called Hilbert vector space [BRE1] that are fed with a scalar product of functions and have an infinite number of dimensions requiring a criterion of convergence for any vector expansion in any Hilbertian base. We recall the definition of the scalar product of two functions f and g as follows:

$(f,g) = \int_\Omega fg \, d\omega$, Ω being the domain where the functions are not zero in relation to the Hilbertian norm $\|f\|_{L^2} = \sqrt{(f,f)}$.

Let the *linear mapping* A from a scalar function $f(x, y, z, t)$ depending on variables (x, y, z, t) to another function $g(x', y', z', t')$ be considered a real number as regards (x, y, z, t) because it depends only on other variables (x', y', z', t'). So, writing $g = A(f)$, which is classically denoted as $g = <A, f>$ by mathematicians, the *Riesz-Frechet representation theorem* in Hilbert space [BRE1] teaches us that there is a single function $G(x, y, z, t, x', y', z', t')$ depending here on all the previous variables, so that g is represented by the following scalar product of G and f as regards the variables (x, y, z, t):

$$g = \langle A, f \rangle = (G, f) = \int_{\Omega \times T} G(x,y,z,t,x',y',z',t') f(x,y,z,t) dxdydxdt$$

where $(x, y, z) \in \Omega$, $t \in T$.

The function G is called the *Green's function or kernel* tied to the linear mapping A. This is the basic tool of the linear modeling of physical phenomena, and is very pertinent for their numerical simulation which is the computer-aided prediction of their behavior in space and time domains by means of "discretization" and numerical computation.

1.1.3. *Accounting for the "natural symmetries" of physical laws*

We have to understand the *natural symmetry* of physical laws, as this changes after a given geometric transformation that can be a translation, a rotation and a reflection through a plane. The most important symmetry encountered in nature is the symmetry or invariance per translation, or that we call *translation symmetry*.

Let $G(\vec{r},\vec{r}',t,t')$ be a Green's function depending on two points and times in space-time. Processing both a space translation having the vector $\vec{h}$ and a time shift τ, the Green's function $G(\vec{r}+\vec{h},\vec{r}'+\vec{h},t+\tau,t'+\tau)$ has to not depend on the translation vector and time shift in the case of translation symmetry. The result of differentiation with respect to the translation vector and time shift has to be zero:

$$\frac{\partial G}{\partial \vec{h}} = \frac{\partial G}{\partial \vec{r}} + \frac{\partial G}{\partial \vec{r}'} = 0$$

$$\frac{\partial G}{\partial \tau} = \frac{\partial G}{\partial t} + \frac{\partial G}{\partial t'} = 0$$

4   Electromagnetism and Interconnections

This gives the single solution $G = G(\vec{r} - \vec{r}', t - t')$ which is called a *convolution kernel*. Then, the scalar product (G, f) becomes the *convolution product* denoted as $G * f$, which is the basic tool of linear systems modeling in so far as they have translation symmetry.

Other symmetries are often encountered when considering physical phenomena such as electromagnetism. The most important after the translation symmetries is the one tied to the orientation of space having to be included into the relations between electromagnetic features. It will be seen that this symmetry completed by the symmetry of invariance per rotation (*isotropy*) contributes to obtaining the first couple of Maxwell's equations.

While a symmetry cannot be broken inside an equation, which means the two parts of it have to have the same symmetry, it can be broken outside it, thus leading to a new equation: this is the case in the second couple of Maxwell's equations concerning the electromagnetic behavior of anisotropic materials, thus leaving the symmetry tied to the *isotropy* (invariance per rotation). In this case, the scalar relations have to be replaced by tensor ones.

A final symmetry concerns reciprocal linear relations: if g is a square integrable function linearly tied to another one f through the mapping A, then f is linearly tied to g through the reverse mapping A^{-1} which should lead to a new Green's function G' according to the Riesz-Frechet theorem. This is written as follows:

$$g = \langle A, f \rangle = (G, f) \Leftrightarrow f = \langle A^{-1}, g \rangle = (G', f)$$

This then becomes

$$g = \langle A, \langle A^{-1}, g \rangle \rangle = (G, (G', g))$$

whatever g is. The last equation is linear, so it is relevant to apply the properties of the mathematical space L^1. The Riesz-Frechet theorem invites us to consider the existence of a new Green's kernel N, so that $g = (N, g)$. The properties of N, being non-compatible with those of the square integrable functions, have to be understood only in the light of the theory of distributions, on which we now focus.

1.1.4. *Motivation for using the theory of distributions*

In order to satisfy in a full and efficient way the needs of mathematical modeling of electromagnetic phenomena, we are led to use the language arising from the theory of distributions formulated by L. Schwarz in 1950 [SCH]. This theoretical approach allows us to avoid the problems related to managing the discontinuities of electromagnetic features and also to highlight the basic differential relations between them.

It can be introduced according to the following:

– the electromagnetic features show rough variations through boundaries that can be either closed curves or surfaces;

– then, some measurement file can be modeled by one or several functions that are necessarily continuously derivable inside and outside a given domain and discontinuous and so not derivable on its boundary. It also happens for the requirements of point sources modeling that these functions are not square integrable;

– this is easily modeled using the theory of distributions and the use of the known Dirac distribution that we consider later on;

– furthermore, the order and the support of a distribution give a powerful means for shedding new light on why and how there are differential relations in Maxwell's equations (section 1.3) between electromagnetic features;

– in conclusion, the theory of distributions saves the "natural symmetry" between any linear relation and its reverse.

1.1.5. *Quick review of the theory of distributions*

1.1.5.1. *Definition of a distribution*

A *distribution* is a linear mapping from the set of indefinitely derivable functions, which is a vector space denoted as $D(\Re^n)$, having a bounded support defined inside $\Re^n$ to the set of real or complex numbers.

Then, the image of any function φ through the distribution T is the number $T(\varphi)$ that is denoted as

$$\langle T, \underline{\varphi} \rangle \tag{1.1}$$

according to the theory of distributions. This number has to be physically understood as the result of a measurement. It is declared $T = 0$ if and only if $\langle T, \underline{\varphi} \rangle = 0$. Let us recall the definition of a linear mapping as follows:

$$\langle T, \alpha\varphi_1 + \beta\varphi_2 \rangle = \alpha \langle T, \varphi_2 \rangle + \beta \langle T, \varphi_2 \rangle, \quad \forall (\alpha, \beta) \in \Re \times \Re$$

This has to be completed by a criterion of convergence: m being an integer, T_m converges towards T if and only if $<T_m, u>$ converges towards $<T, u>$ whatever u belonging to $D(\Re^n)$.

1.1.5.2. *Support of a distribution*

The support of a *distribution* is the complementary subset of the subset of all the points where it is zero according to the theory of distributions as defined above. It is a powerful concept for setting the foundations of electromagnetism (section 1.3.3).

1.1.5.3. *Regular distributions*

Let f be an integrable function in the sense of Lebesgue within a bounded area. It can be defined by the distribution T_f so that

$$\forall \underline{\varphi}(M) \in D, \quad \langle T_f, \underline{\varphi} \rangle = \iiint_{M \in V} f(M) \underline{\varphi}(M) \, dv \quad [1.2]$$

This distribution is related to the set of all the functions equal to $\underline{f}$ almost everywhere: it is called a *regular* distribution. Physicists can confuse these functions with distributions and they write usually f instead of T_f. However, it is necessary to know that it is identifying a function with a mapping from the set to which this function belongs to the set of numbers, in terms of simplifying the writing. An example in electromagnetism is shown in section 1.1.6.

1.1.5.4. *Singular distributions*

Any distribution that cannot be considered as regular is called *singular*. The most famous singular distribution is the *Dirac distribution* δ_a at the point a defined as follows:

$$\text{into } \Re^n: \forall \underline{\varphi} \in D, \quad \langle \delta_a, \underline{\varphi} \rangle = \underline{\varphi}(a) \quad [1.3]$$

This rigorous definition is applied in formulae [2.16] (sections 2.2.3 and 4.2.8) and [4.12] (section 4.1.4). Physicists [JAC2] write:

$$\text{into } \Re: \delta_a = \delta(x-a) \quad \text{and} \quad \int_{x_1}^{x_2} \varphi(x) \delta(x-a) \, dx = \varphi(a), \quad x_1 < a < x_2$$

This is used in spectral analysis (section 4.1.4), modal analysis (section 4.1.5) and Foucault's current analysis (section 4.3.1).

Here, again, the following expressions are not considered rigorous by mathematicians, but they are frequently used anyway in the scientific literature.

On a surface S into $\Re^3$, the distribution δ_S is defined by

$$\forall \underline{\varphi} \in \mathsf{D}, \quad \iiint \underline{\alpha} \delta_S \varphi \, \mathrm{d}v = \iint_S \underline{\alpha} \varphi \, \mathrm{d}S \quad [1.4]$$

This distribution is needed for the method of moments (section 2.1.3) and harmonic wave analysis (section 2.2.3).

On a curve C into $\Re^3$, the distribution δ_C is defined by

$$\forall \underline{\varphi} \in \mathsf{D}, \quad \iiint \delta_C \varphi \, \mathrm{d}v = \int_C \underline{\varphi} \, \mathrm{d}C \quad [1.5]$$

This distribution is needed for electrostatic modeling in section 4.2.6 (formula [4.60]) and in TEM numerical analysis in section 4.2.8.

1.1.5.5. *Finite order distribution*

A distribution T is said to have *order k* if there exists a positive real number C and an integer $k > 0$ so that $\forall \varphi \in D$:

$$|\langle T, \varphi \rangle| \le C \sum_{\alpha_1 + \alpha_2 + \cdots \le k} \sup \left| \frac{\partial^{\alpha_1 + \alpha_2 + \cdots}}{\partial x_1^{\alpha_1} \partial x_2^{\alpha_2} \cdots} \varphi \right|$$

where sup is the upper bound of the domain defining T.

A distribution having an order not greater than k is always a mapping from the set of functions k times continuously differentiable with bounded support to the set of numbers (real or complex). First-order distributions ($k = 1$) are required for the foundations of electromagnetism (section 1.3.3).

1.1.5.6. *Kernel distribution of a linear mapping, "elementary solution" of an equation*

Let us consider a linear mapping A from the set $D(R^m)$ of the functions indefinitely differentiable defined in R^m as φ to the set $D'(R^n)$ of those defined as $\Psi = A(\varphi)$.

According to the *Schwarz theorem of kernels*, any linear mapping A from $D(R^m)$ to $D'(R^n)$ defines a true new distribution V belonging to $D'(R^m \times R^n)$, called the *Green's kernel distribution of linear mapping*, denoted as $N(A)$, so that

$$\langle V, \varphi \rangle = \Psi \Rightarrow \exists A : \begin{cases} \Psi = A(\varphi) \\ V = N(A) \end{cases}$$

This is a basic result we will use later on for demonstrating the basic equations of electromagnetism:

$$\Psi = A(\varphi) \Rightarrow \Psi = \langle N(A), \varphi \rangle$$

Coming back to "physical" expressions, we can write

$$\int \varphi(x') \delta(x - x') \, dx' = \varphi(x)$$

where x' belongs to a domain in R^m. The convolution product can replace this, written rigorously as $\varphi = \varphi * \delta$. Let us compute $A(\varphi)$:

$$A(\varphi)(x) = A\left[\int \varphi(x') \delta(x - x') \, dx' \right] = \int \varphi(x') A[\delta(x - x')] \, dx'$$

because of the linearity of A as regards any convolution product.

Then, we obtain the generalized Green's kernel related to the linear mapping A as the image of the Dirac distribution through A:

$$A[\delta] = \langle N(A), \delta \rangle = G$$

This is an extension of that already presented concerning the Green's kernels tied to the linear mappings into Hilbert space. This leads to

$$A^{-1}(G) = \delta$$

Then, G is said to be an *elementary solution* of the equation $A^{-1}(u) = \delta$. Knowing this elementary solution, the solution of the equation $A^{-1}(u) = f$ can be obtained if there is a linear mapping B commuting with A^{-1} so that $f = B(\delta)$. Then, $B[A^{-1}(G)] = A^{-1}[B(G)] = f$ which means the solution u is equal to $B(G)$. In so far as A^{-1} commutes with the translations that means the physical law has a "symmetry translation", B can be chosen as the linear operator of the convolution product of the function f by the Dirac distribution, then the solution becomes, in this particular but basic case, a convolution product of f by the Green's kernel $u = f * G$. The elementary solutions will be frequently used in sections 2.2.3 and 4.2.8.

To finish, let us have a look at the symmetry tied to the reverse linear relation corresponding to A^{-1}. This is linked to the Green's function G', so that

$$A(G') = \delta = A^{-1}(G) \Rightarrow G' = A^{-2}(G)$$

1.1.5.7. *Vector, tensor and pseudo-tensor distributions*

For applications to electromagnetism, it is necessary to extend the range of the previous discussion. For instance, any mapping from a set of functions defined in $\Re^m$ to a set of n numbers belonging to $\Re^n$ is a *vector distribution*.

Any mapping being the kernel of a linear mapping from a vector distribution to another vector distribution is a *tensor distribution* having order 2.

When a tensor distribution is completely non-symmetric, its twice contracted product with a vector distribution (that is, a classical product of a matrix and a vector on its right side) leads to what is called a *pseudo-vector distribution*. Physicists [JAC3] call them *axial vector distributions* in contrast to the standard vector distributions called *polar vector distributions*.

Axial and *polar* vector distributions are basic mathematical entities used in electromagnetism, as is discussed in section 1.1.6.

1.1.6. *Application to electromagnetism*

As mentioned at the beginning of this chapter, electromagnetic phenomena are basically linear. Moreover, it is recalled that some physical factors involved in measurements in electromagnetism, such as a magnetic field, can be non-zero only within a bounded part of space while others remain zero, such as electrical current density. According to the theory of distributions, this property can be used to characterize accurately the relations between these factors.

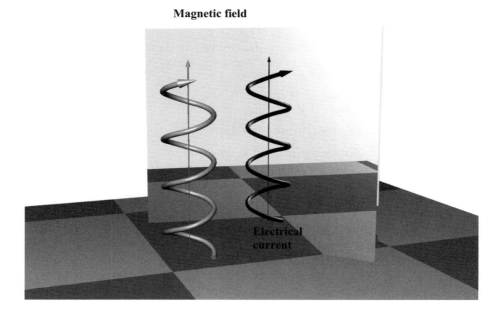

Figure 1.1. *Change of space orientation by reflection through a mirror of a coil through which an electrical current travels and corresponding magnetic field*

These relations have to be independent of the choice of space and time origin and should not change during rotation of the reference axis. Nevertheless, these relations depend basically on the *space orientation* [JAC4], that is, the choice of a positive direction of rotation around any oriented axis in space according the famous corkscrew law. Indeed, this is the most important feature characterizing electromagnetic phenomena like electrical induction in a magnetic coil or magnetization around an electrical wire as shown by classic experiments usually taught in school. Here, the reader has to keep in mind that the results of such experiments are reversed since they are observed after reflection in a mirror that corresponds geometricly to symmetry through a plane. As an example, Figure 1.1 shows a magnetic field created by an electrical current traveling along a coil and its reflection through a mirror so showing the change of the direction of the electrical current rotation.

This dependence of the relations in electromagnetism on space orientation leads to the formal classification of the basic electromagnetic features from which any other can be deduced:

Theoretical Foundations of Electromagnetism   11

– The *polar vector distributions*, having their three coordinates independent of the space orientation. They are:

- *electrical field:* $T_{\underline{\vec{E}}}$, identified "almost everywhere" (outside any parts of space having zero volume like surfaces, curves or stand-alone points) with a vector function $\vec{E}$ that is once continuously differentiable;

- *electrical current density:* $T_{\underline{\vec{j}}}$, identified "almost everywhere" with a continuous vector function $\vec{j}$ not necessarily differentiable.

– The *axial pseudo-vector distributions*, having their three coordinates depending on the space orientation; then the sign of their coordinates changes in the case of symmetry through a plane. They are:

- *magnetic field:* $T_{\underline{\vec{H}}}$, identified "almost everywhere" with a vector function $\underline{\vec{H}}$ that is once continuously differentiable;

- *magnetic current density:* $T_{\underline{\vec{j}_m}}$, identified "almost everywhere" with a continuous vector function $\underline{\vec{j}}_m$. This is only a theoretical concept since there is no elementary particle related to a magnetic current, in contrast to electrons that are the basis of electrical current. This concept is very useful, however, for writing the general equations of electromagnetic interactions through the free space between conductors [JAC5].

1.2. Vector analysis review according to the theory of distributions

1.2.1. *Derivation of discontinuous functions defined on R*

Let $\underline{f}(x)$ be an integrable continuous function which is differentiable outside the point $x = a$. Let T_f be the distribution related to f.

The definition of the first derivative $(T_f)'$ of T_f according to the formula $\langle (T_f)', \varphi \rangle = -\langle T_f, \varphi' \rangle$ and relation [1.3] defining δ_a lead to the classical result

$$(T_{\underline{f}})' = T_{\underline{f}'} + \underline{\sigma}_f \delta_a \quad \text{with} \quad \underline{\sigma}_f = \underline{f}(a^+) - \underline{f}(a^-) \qquad [1.6]$$

where $T_{\underline{f}'}$ is the distribution related to the function $\underline{f}'(x)$, the derivative of $\underline{f}(x)$, assumed to be locally integrable.

In order to simplify the expressions, the *derivative according to the theory of distributions* is denoted f' instead of $(T_f)'$, while the derivative according to the classical analysis of functions is denoted $\{f'\}$ instead of $T_{f'}$. Relation [1.6] becomes

$$\underline{f}' = \{\underline{f}'\} + \underline{\sigma}_f \delta(x-a) \qquad [1.7]$$

with $\delta(x-a)$ instead of δ_a. This is used in section 2.1.1.

1.2.2. Derivative of linear mappings

The definition of the *derivative of any distribution or linear mapping* is based on the formula

$$\langle (T_f)', \varphi \rangle = -\langle T_f, \varphi' \rangle$$

This formula and relation [1.3] lead to the definition of the first-order derivative of the Dirac distribution:

$$\langle \delta'_a, \varphi \rangle = -\varphi'(a)$$

Remembering the definition of the order of a distribution, the order of δ'_a is not greater than 1, while the order of δ_a is zero. Thus, it is shown that the derivative operator is related to the first-order distribution δ'_a in terms of the Schwarz theorem of kernels.

For establishing the foundations of electromagnetism (section 1.3.3), we shall require that any linear mapping from the set of once differentiable functions defined in $D^1(\Re^m)$ to $\Re$ can be written as

$$\sum_{n=1}^{n=m} \Re_n(x_1, x_2, \ldots x_m) \delta'(x_n)$$

1.2.3. Derivation of discontinuous functions on a surface in $\Re^3$

Let $f(x, y, z)$ be a locally integrable function that is differentiable inside the complementary set of a surface S inside $\Re^3$. It defines a distribution T_f. $\vec{n}(n_x, n_y, n_z)$ being the normal to S assumed to be orientable, the calculus of the partial derivative along the x axis (similarly along the other y or z axes) leads to

$$\frac{\partial \underline{f}}{\partial x} = \left\{\frac{\partial f}{\partial x}\right\} + n_x \underline{\sigma}_f \delta_S \quad [1.8]$$

We have to note that $\underline{\sigma}_f$ is the jump of f through S in the direction of $\vec{n}$. The product $n_x \underline{\sigma}_f$ is independent of this direction. Let $\vec{\nabla} T_f$ be the vector distribution, the coordinates of which are $\partial T_f / \partial x$, $\partial T_f / \partial y$, $\partial T_f / \partial z$ (partial derivatives according to the theory of distributions). This leads to the following simplified formula:

$$\vec{\nabla} \underline{f} = \{\vec{\nabla} \underline{f}\} + \vec{n} \underline{\sigma}_f \delta_S \quad [1.9]$$

The differential operator ∇ is called *nabla*.

1.2.4. Derivation of vector distributions in $\mathfrak{R}^3$

Let a set of three distributions T_{A_x}, T_{A_y}, T_{A_z} define the vector distribution $T_{\vec{A}}$ related to the vector $\vec{A}(\underline{A}_x, \underline{A}_y, \underline{A}_z)$ derivable in the complementary set of an oriented surface S having a normal $\vec{n}(n_x, n_y, n_z)$ at any point.

We get the following formula again showing the simplified expressions needed for processing the discontinuities in electromagnetism:

$$\begin{aligned} \vec{\nabla} \cdot \underline{\vec{A}} &= \{\vec{\nabla} \cdot \underline{\vec{A}}\} + \vec{n} \cdot \underline{\sigma}_{\vec{A}} \delta_S \\ \vec{\nabla} \wedge \underline{\vec{A}} &= \{\vec{\nabla} \wedge \underline{\vec{A}}\} + \vec{n} \wedge \underline{\sigma}_{\vec{A}} \delta_S \end{aligned} \quad [1.10]$$

with $\underline{\sigma}_{\vec{A}}(\underline{\sigma}_{A_x}, \underline{\sigma}_{A_y}, \underline{\sigma}_{A_z})$ defining the jump of the vector $\vec{A}$ through S, the coordinates of which are equal to the jump of the functions $\underline{A}_x$, $\underline{A}_y$, $\underline{A}_z$.

1.2.5. Algebra of the operator ∇

Let us recall that the nabla operator [ANG] has the properties of both vectors and of differential operators. The latter are based on the rule of differentiation of a product of functions or distributions. The following are useful formulae (to be understood in the framework of the theory of distributions):

14   Electromagnetism and Interconnections

$$\vec{\nabla}(fg) = f\vec{\nabla}g + g\vec{\nabla}f$$
$$\vec{\nabla} \wedge (\vec{\nabla}f) = \vec{0}$$
$$\vec{\nabla}(\lambda \vec{U}) = \vec{U}(\vec{\nabla}\lambda) + \lambda(\vec{\nabla}\vec{U})$$
$$\vec{U}(\vec{\nabla}\lambda) = (\vec{U}\vec{\nabla})\lambda$$
$$\vec{\nabla} \wedge (\lambda \vec{U}) = (\vec{\nabla}\lambda) \wedge \vec{U} + \lambda(\vec{\nabla} \wedge \vec{U})$$
$$\vec{\nabla}(\vec{U} \wedge \vec{V}) = \vec{V}(\vec{\nabla} \wedge \vec{U}) - \vec{U}(\vec{\nabla} \wedge \vec{V})$$
$$\vec{\nabla}(\vec{\nabla} \wedge \vec{U}) = 0$$
$$\vec{\nabla}(\vec{\nabla}f) = \nabla^2 f$$
$$\vec{\nabla} \wedge (\vec{\nabla} \wedge \vec{U}) = \vec{\nabla}(\vec{\nabla}\vec{U}) - \nabla^2 \vec{U}$$
$$\vec{\nabla} \wedge (\vec{U} \wedge \vec{V}) = \vec{U}(\vec{\nabla}\vec{V}) - \vec{V}(\vec{\nabla}\vec{U}) + (\vec{V}\vec{\nabla})\vec{U} - (\vec{U}\vec{\nabla})\vec{V}$$
$$\vec{\nabla}(\vec{U}\vec{V}) = \vec{U} \wedge (\vec{\nabla} \wedge \vec{V}) + \vec{V} \wedge (\vec{\nabla} \wedge \vec{U}) + (\vec{V}\vec{\nabla})\vec{U} + (\vec{U}\vec{\nabla})\vec{V}$$

1.3. Maxwell's equations according to the theory of distributions

1.3.1. *Symmetries and duality in electromagnetism*

According to the introduction of this chapter, we deal with the natural symmetries arising from experiments in order to derive the well-known Maxwell's equations [FOU]. Then, we are led to consider two levels of symmetries as laws about the whole space: the first level contains the symmetries tied to the properties of Euclidian space free of condensed matter while the second level is related to the electromagnetic behavior of condensed matter.

The most important symmetry is the *duality* about which any relationship between electrical and/or magnetic features has a corresponding relationship between magnetic and/or electrical features.

1.3.2. *The symmetry laws of distributions in electromagnetism*

Any "axial" vector distribution of magnetic field $\underline{\tilde{H}}$ (respectively "polar" vector distribution of electrical field $\underline{\vec{E}}$) is linearly related to a "polar" vector distribution of electrical current density $\underline{\vec{j}}$ (respectively to an "axial" vector distribution of magnetic current density $\underline{\tilde{j}}_m$) independent of any origin in the space and in an isotropic way (invariant by rotation). The constraints are that since the field $\underline{\tilde{H}}$

(respectively $\vec{E}$) is uniform inside a bounded area, then the current $\vec{j} = \vec{0}$ (respectively $\vec{\underline{j}}_m = \vec{0}$).

1.3.3. *Application to the first couple of Maxwell's equations*

The previous basic law of electromagnetism shows an efficient way to obtain Maxwell's equations from geometric features only thanks to the powerful help of the Schwarz theory of distributions.

According to the previous basic law of electromagnetism, the linear mapping R (respectively R_m), so that $\vec{j} = R(\vec{\underline{H}})$ (respectively $\vec{\underline{j}}_m = R_m(\vec{E})$) defines a linear operator.

Furthermore, the linear operator R (respectively R_m) is related to the kernel $N(R)$ of the linear mapping R:

$$\vec{\underline{H}} \text{ (respectively } \vec{E}) \xrightarrow{R} \vec{j} \text{ (respectively } \vec{\underline{j}}_m)$$

which can be identified, according to the Schwarz theorem of kernels, as a distribution.

In the space domain where there is a non-zero uniform field, the corresponding current is zero (according to the basic law); the reciprocal is not true, which shows the mapping R (respectively R_m) reduces the support of the corresponding distribution. Therefore, according to another *Schwarz theorem* for which "any linear operator reducing the *support of a distribution* is necessarily a differential operator", the linear operators involved are necessarily differential and are a linear combination of partial derivatives versus coordinates.

Since the linear mapping operates on $\vec{\underline{H}}$ (respectively $\vec{E}$) which is identified as a first-order differentiable function "almost everywhere", the corresponding distribution has an order not greater than 1, which means the linear combination of partial derivatives representing R can only contain first-order partial derivatives. This is rigorously expressed as follows, in the framework of the theory of distributions, following the previous discussion:

$$\vec{j} = R(\vec{\underline{H}}) = \left\langle N(R), \vec{\underline{H}} \right\rangle = \left\langle \sum_{n=1}^{n=3} R_n(x_1, x_2, x_3) \delta'(x_n), \vec{\underline{H}} \right\rangle = \sum_{n=1}^{n=3} R_n(x_1, x_2, x_3) \frac{\partial}{\partial x_n}(\vec{\underline{H}}) \quad [1.11]$$

16 Electromagnetism and Interconnections

Each R_n is a second-order tensor equivalent to a square matrix having elements denoted R_{lmn}, n being fixed, thus allowing us to compute the coordinates $\underline{j}_l$ of the current density $\vec{j}$ versus the coordinates $\underline{H}_m$ of the magnetic field $\vec{\underline{H}}$ as follows:

$$\underline{j}_l = \sum_{m=1}^{m=3}\sum_{n=1}^{n=3} R_{lmn}(x_1, x_2, x_3)\frac{\partial H_n}{\partial x_n}$$

Thus, the set of scalar functions R_{lmn}, n now being a "variable" integer, becomes a third-order tensor from which we can obtain interesting relations due to the geometric properties: these do not depend on the origin of the coordinates (*translation symmetry*), isotropy, but do depend on the space orientation.

Let us apply first *translation symmetry*. This means the laws $\vec{j} = R(\vec{\underline{H}})$ or $\vec{\underline{j}}_m = R_m(\vec{E})$ are invariant through any translation.

Let T be a translation operator working simultaneously on $\vec{j}$ and $\vec{\underline{H}}$ (or $\vec{\underline{j}}_m$ and $\vec{E}$). We set

$$\vec{j}' = T\vec{j}, \quad \vec{\underline{H}}' = T\vec{\underline{H}}$$

We again obtain $\vec{j}' = R\vec{\underline{H}}'$ thanks to the invariance through translation. We deduce

$$\vec{j}' = T^{-1}RT(\vec{\underline{H}}')$$

Then $T^{-1}RT = R$, so $RT = TR$; R has to commute with the translations, which gives the following:

$$RT\vec{\underline{H}}(x_1,x_2,x_3) = R\vec{\underline{H}}(x_1+h_1, x_2+h_2, x_3+h_3) = \sum_{n=1}^{n=3} R_n(x_1,x_2,x_3)\frac{\partial}{\partial x_n}(T\vec{\underline{H}})$$

$$TR\vec{\underline{H}}(x_1,x_2,x_3) = T\sum_{n=1}^{n=3} R_n(x_1,x_2,x_3)\frac{\partial \vec{\underline{H}}}{\partial x_n} = \sum_{n=1}^{n=3} R_n(x_1+h_1, x_2+h_2, x_3+h_3)\frac{\partial}{\partial x_n}(T\vec{\underline{H}})$$

These two equations can be identical only if R_{lmn} are constant in space.

Let us now turn to the *orientation of the space*. It is known, according to Euclidean geometry, that there are two classes of orientation that can be exchanged by symmetry through any plane (reflection at a mirror). This means the coordinates of the "axial" pseudo-vector involved change their sign during the operation of reflection while the sign of the "polar" vector does not change. Let $x_n = x_m$ (n and m being equal to 1, 2 or 3). Therefore, before symmetry:

$$\underline{j}_l = \sum_{m=1}^{m=3}\sum_{n=1}^{n=3} R_{lmn} \frac{\partial \underline{H}_m}{\partial x_n} \qquad [1.12]$$

and after symmetry:

$$\underline{j}_l = \sum_{m=1}^{m=3}\sum_{n=1}^{n=3} R_{lmn} \frac{\partial(-\underline{H}_n)}{\partial x_m} = -\sum_{m=1}^{m=3}\sum_{n=1}^{n=3} R_{lnm} \frac{\partial \underline{H}_m}{\partial x_n} \qquad [1.13]$$

The second members of these two last equations have to be equal whatever the field distribution in space. This can be achieved only if

$$R_{lmn} + R_{lnm} = 0$$

In the same way, another symmetry operating through the new plane $x_l = x_m$ leads to

$$\underline{j}_m = \sum_{l=1}^{l=3}\sum_{n=1}^{n=3} R_{lmn} \frac{\partial(-\underline{H}_l)}{\partial x_n}$$

that is to say

$$\underline{j}_l = \sum_{m=1}^{m=3}\sum_{n=1}^{n=3} R_{mln} \frac{\partial \underline{H}_m}{\partial x_n}$$

Comparing with the situation before symmetry, we obtain $R_{lmn} + R_{lnm} = 0$.

From the above relations, we can write $R_{nml} = -R_{mnl} = R_{mln} = -R_{lmn}$, then $R_{lln} = R_{lmm} = R_{lml} = 0$. So, the third-order tensor R_{lmn} is totally antisymmetric.

Among its $3^3 = 27$ elements, there are 3 that have three equal index numbers and $3 \times 3 \times 2 = 18$ that have two equal index numbers, and thus there are 21 zero elements. There remain 6 non-zero elements, among which 3 have a positive sign, the others having a negative sign and being the opposite of the previous ones.

Thus, R_{lmn} depend only on 3 numbers thanks to behavior of the fields in relation to the orientation of space.

Let us choose them as R_{123}, R_{231}, and R_{312}; we can now write relation [1.12] as follows:

$$\underline{j}_1 = R_{123}\left(\frac{\partial \underline{H}_2}{\partial x_3} - \frac{\partial \underline{H}_3}{\partial x_2}\right)$$

$$\underline{j}_2 = R_{231}\left(\frac{\partial \underline{H}_3}{\partial x_1} - \frac{\partial \underline{H}_1}{\partial x_3}\right) \qquad [1.14]$$

$$\underline{j}_3 = R_{312}\left(\frac{\partial \underline{H}_1}{\partial x_2} - \frac{\partial \underline{H}_2}{\partial x_1}\right)$$

We have to conclude by means of *isotropy* that any rotation of the coordinate axis does not change both the classes of orientation and the projection of the relation $\vec{\underline{j}} = R(\vec{\underline{\underline{H}}})$ on the axis.

Applying the rotation having an axis defined by the equations $x_1 = x_2 = x_3$ and an angle equal to 120°, we carry out the circular permutations $\{1, 2, 3\} \rightarrow \{2, 3, 1\} \rightarrow (3, 2, 1)$ without changing the field coordinates.

Then, we obtain $R_{123} = R_{231} = R_{312}$ that defines a scalar feature. The relation $\vec{\underline{j}} = R(\vec{\underline{\underline{H}}})$ (respectively $\vec{\underline{j}}_m = R_m(\vec{\underline{E}})$) becomes

$$\vec{\underline{j}} = \vec{\nabla} \wedge \vec{\underline{\underline{H}}} \quad \text{respectively} \quad \vec{\underline{j}}_m = \vec{\nabla} \wedge \vec{\underline{E}} \qquad [1.15]$$

This is the first couple of *Maxwell's equations*. The differential linear operator R (or R_m) becomes the so-called *curl vector* denoted as $\vec{\nabla} \wedge$. The antisymmetric tensor R_{lmn} becomes the so-called *space orientation tensor* ε_{lmn} of which the elements are zero when they have two or three equal index numbers, equal to +1 for any set of index numbers being an even permutation of $\{1, 2, 3\}$, and equal to -1 for any set of index numbers being an odd permutation of $\{1, 2, 3\}$.

1.3.4. Behavior law of materials by means of the theory of distributions

Any "polar" vector distribution of current density $\vec{j}$ (respectively "axial" of magnetic current $\vec{j}_m$) is linearly related to a "polar" vector distribution of electrical field $\vec{E}$ (respectively "axial" of magnetic field $\vec{H}$), independently of any time origin and of the orientation of space, with the constraint $\vec{E}$ (respectively $\vec{H}$) being constant versus time requires that $\vec{j} = \vec{0}$ (respectively $\vec{j}_m = \vec{0}$). The kernel of the linear relation is a distribution having an order less than 1 for any media traveled by electromagnetic waves below optical frequencies [LAN1].

1.3.5. Application to the second couple of Maxwell's equations

We are led to use again the Schwarz theorem because the mapping

$$\vec{E} \xrightarrow{\sigma} \vec{j} = \sigma(\vec{E}) \quad \text{(respectively } \vec{H} \xrightarrow{\sigma_m} \vec{j}_m(\vec{H})) \quad [1.16]$$

defines a linear operator σ (respectively σ_m) which is a distribution having an order at the most equal to 1, and the support of $\vec{E}$ (respectively $\vec{H}$) is reduced with, moreover, the property that the relation involved does not depend on the origin of time.

Therefore, the Taylor expansion of the operator σ (respectively σ_m) can be written as follows:

$$\sigma\left(\frac{\partial}{\partial t}\right) = \sigma + \varepsilon \frac{\partial}{\partial t}$$

$$\sigma_m\left(\frac{\partial}{\partial t}\right) = -\sigma_m - \mu \frac{\partial}{\partial t}$$

where, in the general case of anisotropic heterogenous materials (or "media of electromagnetic field propagation"), ε, μ, σ, and σ_m are second-order tensors, generally depending on the space coordinates. They are called, respectively, dielectric permittivity, magnetic permeability, electrical conductivity and magnetic conductivity tensors. From now on, the reader can consider σ_m is zero in physical applications.

Because the behavior of materials does not depend on space orientation, all these tensors have to be symmetric with respect to the same planes as those used for demonstrating the first couple of Maxwell's equations.

Standard values of parameters ε, μ and σ for some materials used in electronic interconnections materials are given in Appendix A.

We can write

$$\vec{j} = \sigma \underline{\vec{E}} + \varepsilon \frac{\partial \underline{\vec{E}}}{\partial t}$$

$$\vec{j}_m = -\sigma_m \underline{\breve{\vec{H}}} - \mu \frac{\partial \underline{\breve{\vec{H}}}}{\partial t}$$

so that the first couple of *Maxwell's equations* becomes

$$\breve{\vec{\nabla}} \wedge \underline{\breve{\vec{H}}} = \sigma \underline{\vec{E}} + \varepsilon \frac{\partial \underline{\vec{E}}}{\partial t} \qquad [1.17]$$

$$\breve{\vec{\nabla}} \wedge \underline{\vec{E}} = -\sigma_m \underline{\breve{\vec{H}}} - \mu \varepsilon \frac{\partial \underline{\breve{\vec{H}}}}{\partial t} \qquad [1.18]$$

1.3.6. *Charge density, current density, continuity equations*

As

$$\vec{\nabla}(\breve{\vec{\nabla}} \wedge \underline{\breve{\vec{H}}}) = 0 \quad \text{(respectively } \vec{\nabla}(\breve{\vec{\nabla}} \wedge \underline{\vec{E}}) = \breve{0})$$

then

$$\vec{\nabla}(\sigma \underline{\vec{E}}) + \vec{\nabla}\left(\varepsilon \frac{\partial \underline{\vec{E}}}{\partial t}\right) = \vec{\nabla}(\sigma \underline{\vec{E}}) + \frac{\partial}{\partial t}\left[\vec{\nabla}(\varepsilon \underline{\vec{E}})\right] = 0$$

$$\left(\text{respectively } \vec{\nabla}(\sigma_m \underline{\breve{\vec{H}}}) + \frac{\partial}{\partial t}\left[\vec{\nabla}(\mu \underline{\breve{\vec{H}}})\right] = \breve{0}\right)$$

The entity $\vec{\nabla}(\varepsilon \underline{\vec{E}}) = \rho$ (respectively $\vec{\nabla}(\mu \underline{\breve{\vec{H}}}) = \breve{\rho}_m$) is a scalar (respectively pseudo-scalar) distribution called the *electrical (respectively magnetic) charge*. The vector distribution $\varepsilon \underline{\vec{E}}$ is called the *electrical induction* $\vec{D}$.

Then, we encounter the so-called *Poisson's equation*:

$$\vec{\nabla}\underline{\vec{D}} = \underline{\rho} \qquad [1.19]$$

In the same way:

$$\vec{\nabla}(\vec{\nabla}\wedge\underline{\vec{E}}) = 0 \quad \text{leads to} \quad \vec{\nabla}(\sigma_m\underline{\breve{\vec{H}}}) + \frac{\partial}{\partial t}\left[\vec{\nabla}(\mu\underline{\breve{\vec{H}}})\right] = 0$$

The pseudo-vector distribution $\mu\underline{\breve{\vec{H}}}$ is called the magnetic induction $\underline{\vec{B}}$. If the initial conditions are zero, it becomes

$$\vec{\nabla}\underline{\breve{\vec{B}}} = \underline{\breve{\rho}}_m \quad \text{(pseudo-scalar)} \qquad [1.20]$$

The second couple of Maxwell's equations consists of equations [1.19] and [1.20].

The vector distribution $\sigma\underline{\vec{E}}$ is called the *conduction electrical current density* $\underline{j}_c$.

From the previous definitions, we obtain the *continuity equation* related to the electrical current:

$$\vec{\nabla}\underline{\vec{j}}_c + \frac{\partial\rho}{\partial t} = 0 \qquad [1.21]$$

The factor $\sigma_m\underline{\breve{\vec{H}}}$ is similar to a "conductor" current density $\underline{\breve{j}}_{cm}$, so that we obtain

$$\vec{\nabla}\underline{\breve{\vec{j}}}_{cm} + \frac{\partial\breve{\rho}_m}{\partial t} = \breve{0} \qquad [1.22]$$

which is the continuity equation related to the magnetic current.

Now, we deal with the third and very classical continuity equation according to the following:

$$\underline{\vec{j}}\,\underline{\vec{E}} + \underline{\breve{\vec{j}}}_m\,\underline{\breve{\vec{H}}} = \sigma E^2 + \sigma_m H^2 + \frac{1}{2}\frac{\partial}{\partial t}(\varepsilon E^2 + \mu H^2)$$

$$= \underline{\vec{E}}(\vec{\nabla}\wedge\underline{\breve{\vec{H}}}) - \underline{\breve{\vec{H}}}(\vec{\nabla}\wedge\underline{\vec{E}}) = \vec{\nabla}(\underline{\breve{\vec{H}}}\wedge\underline{\vec{E}})$$

where the formula $\vec{\nabla}(\vec{a} \wedge \vec{b}) = \vec{b}(\vec{\nabla} \wedge \vec{a}) - \vec{a}(\vec{\nabla} \wedge \vec{b})$ has been applied.

The polar vector distribution $\vec{E} \wedge \vec{\tilde{H}}$ is the density of *electromagnetic power flux* or *Poynting vector* $\underline{\vec{j}}_w$, while the scalar distribution $(\varepsilon/2)E^2 + (\mu/2)H^2$ corresponds to the *density of electromagnetic energy* $\underline{\rho}_w$ and $\sigma \underline{E}^2 + \sigma_m \underline{H}^2$ represents the *dissipated power P* so that

$$\vec{\nabla} \underline{\vec{j}}_w + \frac{\partial \underline{\rho}_w}{\partial t} + P = 0 \qquad [1.23]$$

1.3.7. *Integral form of Maxwell's equations*

1.3.7.1. *Ampere's theorem*

Prior to obtaining the integral form of Maxwell's equations, let us define what an *orientable surface* is according to Figure 1.2.

Integrating the two terms of the equation $\vec{\tilde{\nabla}} \wedge \underline{\vec{\tilde{H}}} = \underline{\vec{j}}$ over an orientable surface (S) bounded by a closed curve (C), we use, first, the left-hand term of the *Stoke formula*:

$$\iint_S (\vec{\tilde{\nabla}} \wedge \underline{\vec{\tilde{H}}}) \vec{\tilde{n}} \, dS = \oint_C \underline{\vec{\tilde{H}}} \, d\vec{\tilde{l}} \qquad [1.24]$$

and, second, from the right-hand term, the *electrical current* flowing through (S) is written

$$I = \iint_S \underline{\vec{j}} \vec{\tilde{n}} \, dS \qquad [1.25]$$

Thus, we obtain *Ampere's theorem* which was investigated experimentally in the 1830s and is usually taught as a basic law in classical courses: "the integration of the magnetic field along any closed curve is equal to the whole current crossing any surface set on this curve". It is illustrated in a later chapter (Figure 3.3, section 3.1.2).

Theoretical Foundations of Electromagnetism 23

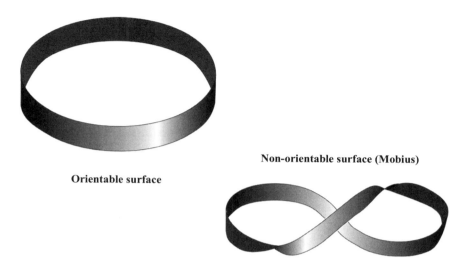

Figure 1.2. *Orientable and non-orientable surfaces*

1.3.7.2. *The Faraday-Lenz theorem*

The equation $\vec{\tilde{\nabla}} \wedge \underline{\vec{\tilde{E}}} = \underline{\vec{\tilde{j}}}_m$ gives an opportunity to use the duality concept in electromagnetism in respect of Ampere's theorem. Indeed, directly applying the Stoke formula to the equation, completed by

$$\underline{\vec{\tilde{j}}}_m = \mu \frac{\partial \vec{\tilde{H}}}{\partial t} = \frac{\partial \vec{\tilde{B}}}{\partial t}$$

we obtain, when $\sigma_m = 0$, the *Faraday–Lenz theorem* according to which "the integration of the electrical field along any closed curve is equal to the derivative of the flux of magnetic induction across any orientable surface set on this curve". This can be illustrated by Figure 1.1, where it is necessary that the magnetic field be time varying, thus generating an induced electrical current.

1.3.7.3. *The Gauss theorem*

We integrate the two terms of the equation $\vec{\nabla}\vec{D} = \rho$ into a bounded area of space. In the first term *Green's formula* is used as follows:

$$\iiint_V \vec{\nabla}\vec{D} \, dV = \oiint_S \vec{D}\vec{n} \, dS \qquad [1.26]$$

while in the second term there is the total electrical charge:

$$Q = \iiint_V \rho \, dV \qquad [1.27]$$

Thus, we obtain the *Gauss theorem* according to which "the flux of the electrical induction across a closed surface is equal to the whole electrical charge included into this surface".

1.4. Conclusion

Maxwell's equations have been demonstrated by means of sharing them into separated equations showing two levels of natural symmetries:

– "strong symmetries", including the duality between electrical and magnetic features, linking axial and polar distributions independently of the material of which the media are made;

– "weak symmetries" linking similar (polar exclusively or axial) distributions and closely depending on the even interconnection material symmetries (which are those of their tensor characteristics [ε], [μ], and [σ]).

These symmetries are the leading thread of the foundations of electromagnetism presented aimed at advanced transmission line modeling in modern interconnection networks.

Chapter 2

Full Wave Analysis

Applying the theory of distributions, the initial and boundary conditions are accounted for automatically which leads to the reflections of the electromagnetic field on conductors and their electromagnetic images. This is completed by presenting the moments for solving Maxwell's equations numerically with respect to the boundary and initial conditions. We deal with the scalar and vector potentials, thus giving the opportunity of pointing out the duality in electromagnetism that is much used later on. Harmonic wave analysis is developed through the use of Green's functions leading to surface integral equations. This is achieved by the topological analysis of the "physically significant" electromagnetic interferences inside any electronic system by means of the theory of oriented graphs in order to obtain the non-zero elements of the matrices involved in numerical computations close to the main diagonal.

2.1. Discontinuities in electromagnetism

2.1.1. *Initial and boundary conditions according to the theory of distributions*

The two couples of Maxwell's equations have been written by means of the theory of distributions. The distributions involved have been considered equal "almost everywhere" (that is to say, here, outside a given surface (S)) to a vector function at least continuous, more accurately strictly continuous if it defines a current and continuously differentiable if it defines a field.

According to the theory of distributions (formulae [1.7] and [1.10]), the linear differential operators can be developed as the sum of two operators, the first one being related to a regular distribution and the second one being related to the singular Dirac distribution δ_S defined on a surface (S)). From this we obtain the first couple of Maxwell's equations:

$$\{\vec{\nabla} \wedge \underline{\vec{H}}\} + \vec{n} \wedge \sigma_{\vec{H}} \delta_S = \sigma \underline{\vec{E}} + \varepsilon \left\{\frac{\partial \underline{\vec{E}}}{\partial t}\right\} + \varepsilon \sigma_{\vec{E}} \delta_{t_0}$$

$$\{\vec{\nabla} \wedge \underline{\vec{E}}\} + \vec{n} \wedge \sigma_{\vec{E}} \delta_S = -\breve{\sigma}_m \underline{\vec{H}} - \mu \left\{\frac{\partial \underline{\vec{H}}}{\partial t}\right\} - \mu \sigma_{\vec{H}} \delta_{t_0}$$

where $\sigma_{\vec{E}} \delta_{t_0}$ and $\sigma_{\vec{H}} \delta_{t_0}$ are the jumps of the distributions $\underline{\vec{E}}$ and $\underline{\vec{H}}$ respectively to the time t_0.

The regular distribution $\{\sigma \underline{\vec{E}}\} = \underline{\vec{j}}_v$ (respectively $\{\breve{\sigma}_m \underline{\vec{H}}\} = \underline{\vec{j}}_{mv}$) defines the *electrical* (respectively *magnetic*) *conduction or Ohm's current bulk density*.

The regular distribution

$$\underline{\vec{j}}_{dv} = \varepsilon \left\{\frac{\partial \underline{\vec{E}}}{\partial t}\right\}$$

defines the *Maxwell or displacement current bulk density*.

The singular distribution $\vec{n} \wedge \sigma_{\vec{H}} \delta_S = \underline{\vec{j}}_S \delta_S$ (respectively $\vec{n} \wedge \sigma_{\vec{E}} \delta_S = \underline{\vec{j}}_{mS} \delta_S$) defines the *electrical* (respectively *magnetic*) *current surface density*:

$$\vec{n} \wedge \sigma_{\vec{H}} = \underline{\vec{j}}_S \qquad [2.1]$$

(respectively $\vec{n} \wedge \sigma_{\vec{E}} = \underline{\vec{j}}_{mS}$). $\qquad [2.2]$

The *electrical* and *magnetic Poisson's equations* are, respectively:

$$\{\vec{\nabla} \underline{\vec{D}}\} + \vec{n} \sigma_{\vec{D}} \delta_S = \underline{\rho}$$
$$\{\vec{\nabla} \underline{\vec{B}}\} + \vec{n} \sigma_{\vec{B}} \delta_S = \breve{\rho}_m$$

The regular distribution $\{\vec{\nabla}\underline{\vec{D}}\} = \underline{\rho}_v$ (respectively $\{\vec{\nabla}\underline{\vec{\breve{B}}}\} = \underline{\breve{\rho}}_{mv}$) defines the *electrical* (respectively *magnetic*) *charge bulk density*. The singular distribution $\vec{n}\sigma_{\underline{\vec{D}}}\delta_S = \underline{\rho}_S\delta_S$ (respectively $\vec{n}\sigma_{\underline{\vec{\breve{B}}}}\delta_S = \underline{\breve{\rho}}_{mS}\delta_S$) defines the *electrical* (respectively *magnetic*) *charge surface density*:

$$\vec{n}\sigma_{\underline{\vec{D}}} = \underline{\rho}_S \qquad [2.3]$$

(respectively $\vec{n}\sigma_{\underline{\vec{\breve{B}}}} = \underline{\breve{\rho}}_{mS}$) [2.4]

The following is the surface continuity equation which links the electrical current surface density to the electrical charge surface density:

$$\vec{\nabla}(\vec{j}_S\delta_S) + \left(\sigma + \varepsilon\frac{\partial}{\partial t}\right)(\rho_S\delta_S) = 0 \qquad [2.5]$$

Sharing the distributions into regular and singular distributions as above leads to two systems of equations: the first is devoted to the regular distributions in the space domain and singular distributions in the time domain that contains the initial conditions; the second is devoted to the singular distributions in the space domain that contains the boundary conditions on a given surface [BOU]. The distributions are summarized in Table 2.1.

Regular distributions in space domain	Singular distributions in space domain	
	Initial conditions	Boundary conditions
$\{\vec{\nabla}\wedge\underline{\vec{\breve{H}}}\} = \underline{\vec{j}}_v + \varepsilon\left\{\dfrac{\partial\underline{\vec{E}}}{\partial t}\right\} +$	$\varepsilon\sigma_{\underline{\vec{E}}}\delta_{t_0}$	$\vec{n}\wedge\sigma_{\underline{\vec{\breve{H}}}} = \underline{\vec{j}}_S$
$\{\vec{\nabla}\wedge\underline{\vec{E}}\} = -\underline{\vec{\breve{j}}}_{mv} - \mu\left\{\dfrac{\partial\underline{\vec{\breve{H}}}}{\partial t}\right\} -$	$\mu\sigma_{\underline{\vec{\breve{H}}}}\delta_{t_0}$	$\vec{n}\wedge\sigma_{\underline{\vec{E}}} = \underline{\vec{\breve{j}}}_{mS}$
$\{\vec{\nabla}\underline{\vec{D}}\} = \underline{\rho}_v$		$\vec{n}\sigma_{\underline{\vec{D}}} = \underline{\rho}_S$
$\{\vec{\nabla}\underline{\vec{\breve{B}}}\} = \underline{\breve{\rho}}_{mv}$		$\vec{n}\sigma_{\underline{\vec{\breve{B}}}} = \underline{\breve{\rho}}_{mS}$

Table 2.1. *Summary of regular and singular distributions*

Without singular distributions in the time domain tied to non-zero initial conditions, singular distributions in the space domain have to exist in order to avoid

2.1.2. Electromagnetic images, incident and reflected fields

According to the boundary conditions discussed above, both the coordinate of the electrical field tangent to the boundary surface and the coordinate of the magnetic field normal to this surface do not vary through it in so far as there are no charges or currents created on it.

In the general case of non-zero currents and charges, the vector coordinate $\vec{n}\sigma_{\vec{B}}$ (respectively $\vec{n}\sigma_{\vec{D}}$) does not change per symmetry with respect to any normal of the tangent plane at any point belonging to the boundary surface and the vector coordinate $\vec{n} \wedge \sigma_{\vec{E}}$ (respectively $\vec{n} \wedge \sigma_{\vec{H}}$) does not change per symmetry through the tangent plane to the boundary surface. Then, it is possible to define two fields $\underline{\vec{E}}'$ and $\underline{\vec{B}}'$ (respectively $\underline{\vec{D}}'$ and $\underline{\vec{H}}'$) in the space free of the boundary surface (S) that give the real boundary conditions when they are gathered with the true fields $\underline{\vec{E}}$ and $\underline{\vec{B}}$ (respectively $\underline{\vec{D}}$ and $\underline{\vec{H}}$). This is highlighted by the example of the magnetic field $\underline{\vec{H}}$ in Figure 2.1, where its symmetric vector is shown through the tangent plane to the boundary surface with the resulting non-zero tangential coordinate and the resulting zero normal coordinate.

When $\underline{\vec{E}}$ and $\underline{\vec{B}}$ (respectively $\underline{\vec{D}}$ and $\underline{\vec{H}}$) arise from sources like $\underline{\vec{j}}_s$ and $\underline{\rho}_s$ (respectively $\underline{\vec{j}}_{mS}$ and $\underline{\rho}_{mS}$), the sources $\underline{\vec{E}}'$ and $\underline{\vec{B}}'$ (respectively $\underline{\vec{D}}'$ and $\underline{\vec{H}}'$) arise from imaginary sources $\underline{\vec{j}}'_s$ and ρ'_s (respectively $\underline{\vec{j}}'_{mS}$ and $\underline{\rho}'_{mS}$), symmetrically set with respect to the previous sources through the tangent plane to the boundary surface. These imaginary sources are called electromagnetic images of the real sources.

The fields $\underline{\vec{E}}$ and $\underline{\vec{B}}$ (respectively $\underline{\vec{D}}$ and $\underline{\vec{H}}$) arising from the sources $\underline{\vec{j}}_s$ and $\underline{\rho}_s$ (respectively $\underline{\vec{j}}_{mS}$ and $\underline{\rho}_{mS}$) are the *incident (electromagnetic) field*, and $\underline{\vec{E}}'$ and $\underline{\vec{B}}'$ (respectively $\underline{\vec{D}}'$ and $\underline{\vec{H}}'$) arising from the imaginary sources $\underline{\vec{j}}'_s$ and ρ'_s (respectively $\underline{\vec{j}}'_{mS}$ and $\underline{\rho}'_{mS}$) are the *reflected field*.

Since the boundary surface (or surface of discontinuity) is *simply connected* (so made of one part only), each incident field related to a single source generates a single reflected field related to the single electromagnetic image of the source. In contrast, when a discontinuity surface consists of several unconnected parts, any incident field generates an infinite number of reflected fields, each of them being related to an electromagnetic image of either the initial source itself or already generated electromagnetic images of this source.

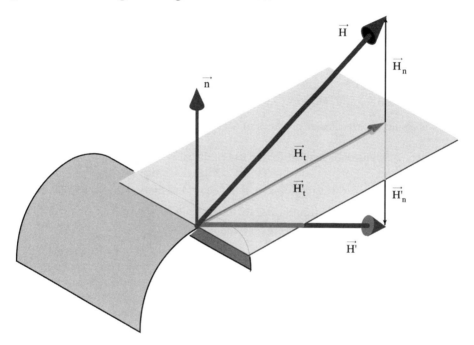

Figure 2.1. *Composition of incident and reflected fields on a boundary surface*

2.1.3. *Method of moments for the numerical computation of electromagnetic fields*

The *method of moments* for solving partial differential equations is widely known nowadays thanks to its numerous applications in electrical engineering developments concerning finite element analysis [BOS] [CHA] [COU] created in the 1950s for the needs of continuous mechanics problems. Its main advantage lies in its accounting automatically for the boundary conditions in which we are interested.

It was mentioned at the beginning of Chapter 1 that the physical features involved can be described by functions belonging to "Hilbert space" inside which a *Hilbertian scalar product of functions* is defined as

$$(f,g) = \iiint_\Omega fg \, dv$$

where Ω is the domain where the physical phenomena are occurring in respect to boundary conditions met on the frontier of this domain denoted $\partial\Omega$.

We need a first extension of this scalar product in the case of vector functions as follows:

$$(\vec{f},\vec{g}) = (f_x,g_x) + (f_y,g_y) + (f_z,g_z)$$

which is a sum of three scalar products of scalar functions.

A second extension is needed for the use of distributions in electromagnetism.

Let $\vec{f}$ be a regular distribution and $\vec{g}$ a vector distribution that is regular almost everywhere except on the frontier $\partial\Omega$ of the domain Ω where it is singular. Then, $\vec{g} = \{\vec{g}\} + \sigma_{\vec{g}}\delta_S$. Note that $\vec{f} = \{\vec{f}\}$ everywhere.

Thus, we can write

$$\iiint_\Omega \vec{f}\vec{g} \, dv = \iiint_\Omega \{\vec{f}\}\{\vec{g}\} \, dv + \iiint_\Omega \{\vec{f}\}\sigma_{\vec{g}}\delta_S \, dv$$

$$\iiint_\Omega \{\vec{f}\}\{\vec{g}\} \, dv = \left(\{\vec{f}\},\{\vec{g}\}\right)$$

which is merely written as $(\vec{f},\vec{g})$ in what follows because $\{\vec{f}\}$, $\{\vec{g}\}$ are regular distributions that can be identified to standard functions assumed to belong to Hilbert space.

Because of the definition of the distribution δ_S (formula [1.4]), we can write

$$\iiint_\Omega \{\vec{f}\}\sigma_{\vec{g}}\delta_S \, dv = \iint_{\partial\Omega} \vec{f}\sigma_{\vec{g}} \, dS$$

Now we deal with the vector analysis formula $\vec{\nabla}(\vec{a}\wedge\vec{b}) = \vec{b}(\vec{\nabla}\wedge\vec{a}) - \vec{a}(\vec{\nabla}\wedge\vec{b})$ that we apply to the case where $\vec{a} = \vec{A}$ is a regular distribution, $\vec{b} = \underline{\breve{\vec{H}}}$.

Integrating the result over the whole domain Ω and applying *Green's formula*

$$\iiint_\Omega \vec{\nabla}\underline{\vec{g}}\,\mathrm{d}v = \iint_{\partial\Omega} \underline{\vec{g}}\vec{n}\,\mathrm{d}S$$

we obtain

$$\iint_{\partial\Omega}(\vec{A}\wedge\underline{\breve{\vec{H}}})\vec{n}\,\mathrm{d}S = (\underline{\breve{\vec{H}}},\vec{\nabla}\wedge\vec{A}) - (\vec{A},\vec{\nabla}\wedge\underline{\breve{\vec{H}}})$$

which shows the mixed product $(\vec{A}\wedge\underline{\breve{\vec{H}}})\vec{n} = \vec{A}(\underline{\breve{\vec{H}}}\wedge\vec{n}) = -\vec{A}\vec{j}_S$.

The first term of the second member $(\underline{\breve{\vec{H}}},\vec{\nabla}\wedge\vec{A})$ contains only a regular distribution identified with functions belonging to "Hilbert space", the magnetic energy being finite. $\vec{A}$ is chosen to be infinitely differentiable.

The second term $(\vec{A},\vec{\nabla}\wedge\underline{\breve{\vec{H}}})$ has to be processed in respect of Maxwell's equations including the boundary and initial conditions:

$$\vec{\nabla}\wedge\underline{\breve{\vec{H}}} = \left\{\vec{\nabla}\wedge\underline{\breve{\vec{H}}}\right\} + \underbrace{\vec{n}\wedge\sigma_{\vec{H}}\delta_S}_{\vec{j}_S} = \sigma\underline{\vec{E}} + \varepsilon\left\{\frac{\partial\vec{E}}{\partial t}\right\} + \varepsilon\sigma_{\vec{E}}\delta_{t_0}$$

Setting the regular distributions in the first member and the singular ones in the second member leads to the first integral equation and, considering duality, we obtain directly the second equation:

$$\left(\vec{\nabla}\wedge\vec{A},\underline{\breve{\vec{H}}}\right) - \left(\vec{A},\sigma\vec{E} + \varepsilon\left\{\frac{\partial\vec{E}}{\partial t}\right\}\right) = -\iint_{\partial\Omega}\vec{j}_S\vec{A}\,\mathrm{d}S + \left(\vec{A},\varepsilon\sigma_{\vec{E}}\delta_{t_0}\right)$$

$$\left(\vec{\nabla}\wedge\vec{A},\underline{\breve{\vec{E}}}\right) - \left(\vec{A}_m,-\mu\frac{\partial\underline{\breve{\vec{H}}}}{\partial t}\right) = -\iint_{\partial\Omega}\vec{j}_{mS}\vec{A}_m\,\mathrm{d}S + \left(\vec{A}_m,\mu\sigma_{\underline{\breve{\vec{H}}}}\delta_{t_0}\right)$$

These two equations are the *method of moments equations*. The arbitrary vector functions $\vec{A}$ and $\vec{A}_m$ are called the *test functions* of the method of moments of solving the partial differential equations. This integral approach of Maxwell's equations leads to the numerical computation of electromagnetic fields. The numerical computation process requires *a method of discretization in the space domain*.

The electromagnetic fields $\vec{\underline{E}}$ and $\vec{\underline{\tilde{H}}}$ are expanded in any base not necessarily "Hilbertian" of which the basic functions $\vec{E}_i(x,y,z)$, $\vec{\tilde{H}}_i(x,y,z)$ are known and their coefficients depending on time $e_i(t)$, $h_i(t)$ have to be computed:

$$\vec{\underline{E}}(x,y,z,t) = \sum_i e_i(t)\vec{E}_i(x,y,z)$$

$$\vec{\underline{\tilde{H}}}(x,y,z,t) = \sum_i h_i(t)\vec{\tilde{H}}_i(x,y,z)$$

Choosing a *finite dimension subspace* of the initial "Hilbert space" where the previous expansions are limited to $i \leq I$, we require a set of I test functions $\vec{A}_{j \leq I}$ and $\vec{A}_{m, j \leq I}$ so that the two integral equations become a set of $2I$ first-order linear differential equations versus time for the $2I$ functions $e_j(t)$ and $h_j(t)$. This can be written in a matrix form using sub-matrices:

$$[M]\frac{\partial}{\partial t}\begin{bmatrix}[e]\\[h]\end{bmatrix} + [N]\begin{bmatrix}[e]\\[h]\end{bmatrix} + [R] = [0] \qquad [2.6]$$

where the transposed matrix of $[e]$ is $[e_1(t), e_2(t),\ldots, e_I(t)]$ and that of $[h]$ is $[h_1(t), h_2(t),\ldots, h_I(t)]$, and

$$[M] = \begin{bmatrix} -[(\varepsilon\vec{A}_j, \vec{E}_i)] & [0] \\ [0] & [(\mu\vec{A}_{mj}, \vec{\tilde{H}}_i)] \end{bmatrix}$$

$$[N] = \begin{bmatrix} -[(\sigma\vec{A}_j, \vec{E}_i)] & [(\vec{\nabla} \wedge \vec{A}_j, \vec{\tilde{H}}_i)] \\ [(\vec{\nabla} \wedge \vec{A}_{mj}, \vec{E}_i)] & [0] \end{bmatrix}$$

$$[R] = \underbrace{\begin{bmatrix}\left[\iint_{\partial\Omega}\vec{j}_S\vec{A}\,dS\right]\\ \left[\iint_{\partial\Omega}\vec{j}_{mS}\vec{A}_m\,dS\right]\end{bmatrix}}_{\text{boundary conditions}} - \underbrace{\begin{bmatrix}[(\vec{A}_j, \varepsilon\sigma_{\vec{E}_i}\delta_{t_0})]\\ -[(\vec{A}_{mj}, \mu\sigma_{\vec{\tilde{H}}_i}\delta_{t_0})]\end{bmatrix}}_{\text{initial conditions}}$$

These equations are well suited for solving problems of electromagnetic pulse propagation inside single waveguides having simple geometries. Nevertheless, they generate too many computations and require too much memory storage capability for any complex network modeling needed for electronic design automation in terms of the capability of modern processors. However, they can be used for modeling small particular parts of networks as detailed in Chapter 6.

Furthermore, the high accuracy of these computations is beyond the available precision of reliable data about the parameters of the materials and the geometry of the usual technological structures inside electronic systems (see Appendix B for geometrical data). This is the reason why we need further realistic but efficient computation methods targeted at high-speed interconnection modeling. Nevertheless, we shall apply such computations in section 6.1.1 for modeling bent and via structures.

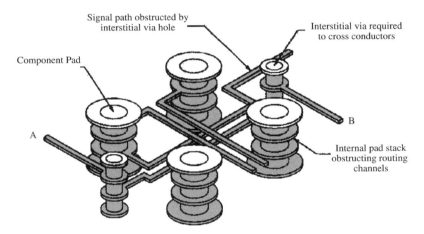

Figure 2.2. *Interconnection network*

2.2. Potentials in electromagnetism

2.2.1. *Scalar and vector potentials, duality between electrical and magnetic potentials*

Let us return to the couples of Maxwell's equations:

34   Electromagnetism and Interconnections

$$\begin{cases} \vec{\nabla} \wedge \underline{\vec{H}} = \sigma \underline{\vec{E}} + \varepsilon \frac{\partial \underline{\vec{E}}}{\partial t} \\ \vec{\nabla} \wedge \underline{\vec{E}} = -\sigma_m \underline{\vec{H}} - \mu \frac{\partial \underline{\vec{H}}}{\partial t} \\ \vec{\nabla}(\varepsilon \underline{\vec{E}}) = \underline{\rho} \\ \vec{\nabla}(\mu \underline{\vec{H}}) = \underline{\rho}_m \end{cases}$$

Because of their linearity, we can share the fields $\underline{\vec{E}}$ and $\underline{\vec{H}}$ as follows:

$$\underline{\vec{E}} = \underline{\vec{E}}_m + \underline{\vec{E}}_e \qquad \underline{\vec{H}} = \underline{\vec{H}}_m + \underline{\vec{H}}_e$$

with the chosen constraints

$$\vec{\nabla}\left(\mu \underline{\vec{H}}_m\right) = \underline{\rho}_m \qquad \vec{\nabla}\left(\varepsilon \underline{\vec{E}}_e\right) = \underline{\rho}$$

from which we obtain

$$\vec{\nabla}\left(\mu \underline{\vec{H}}_e\right) = \vec{0} \qquad \vec{\nabla}\left(\varepsilon \underline{\vec{E}}_m\right) = 0$$

Then, there is a polar vector distribution $\underline{\vec{A}}$ called an *electrical vector potential* and an axial pseudo-vector distribution $\underline{\vec{A}}_m$ called a *magnetic pseudo-vector potential*, so that

$$\mu \underline{\vec{H}}_e = \vec{\nabla} \wedge \underline{\vec{A}} \qquad \varepsilon \underline{\vec{E}}_m = \vec{\nabla} \wedge \underline{\vec{A}}_m \qquad [2.7]$$

because $\vec{\nabla}(\vec{\nabla} \wedge \underline{\vec{A}}) = \vec{0}$ and $\vec{\nabla}(\vec{\nabla} \wedge \underline{\vec{A}}_m) = 0$.

Setting $\underline{\vec{E}} = \underline{\vec{E}}_m + \underline{\vec{E}}_e$ and $\underline{\vec{H}} = \underline{\vec{H}}_m + \underline{\vec{H}}_e$ into

$$\begin{cases} \vec{\nabla} \wedge \underline{\vec{H}} = \sigma \underline{\vec{E}} + \varepsilon \frac{\partial \underline{\vec{E}}}{\partial t} \\ \vec{\nabla} \wedge \underline{\vec{E}} = -\sigma_m \underline{\vec{H}} - \mu \frac{\partial \underline{\vec{H}}}{\partial t} \end{cases}$$

we obtain

$$\tilde{\vec{\nabla}} \wedge \underline{\tilde{\vec{H}}}_m - \sigma \underline{\tilde{\vec{E}}}_m - \varepsilon \frac{\partial \underline{\tilde{\vec{E}}}_m}{\partial t} = \sigma \underline{\tilde{\vec{E}}}_e + \varepsilon \frac{\partial \underline{\tilde{\vec{E}}}_e}{\partial t} - \tilde{\vec{\nabla}} \wedge \underline{\tilde{\vec{H}}}_e \qquad [2.8]$$

$$\tilde{\vec{\nabla}} \wedge \underline{\tilde{\vec{E}}}_m + \sigma_m \underline{\tilde{\vec{H}}}_m + \mu \frac{\partial \underline{\tilde{\vec{H}}}_m}{\partial t} = -\sigma_m \underline{\tilde{\vec{H}}}_e - \mu \frac{\partial \underline{\tilde{\vec{H}}}_e}{\partial t} - \tilde{\vec{\nabla}} \wedge \underline{\tilde{\vec{E}}}_e \qquad [2.9]$$

Here, we require new constraints without lack of generalization. Let us choose to set the parts of the two equations above equal to zero. Using

$$\mu \underline{\tilde{\vec{H}}}_e = \tilde{\vec{\nabla}} \wedge \underline{\vec{A}} \qquad \varepsilon \underline{\tilde{\vec{E}}}_m = \tilde{\vec{\nabla}} \wedge \underline{\tilde{\vec{A}}}_m$$

the first part of [2.8] and the second part of [2.9] become, respectively

$$\tilde{\vec{\nabla}} \wedge \left(\underline{\tilde{\vec{H}}}_m - \frac{\partial \underline{\tilde{\vec{A}}}_m}{\partial t} \right) - \frac{\sigma}{\varepsilon} (\tilde{\vec{\nabla}} \wedge \underline{\tilde{\vec{A}}}_m) = 0 \quad \text{and} \quad \tilde{\vec{\nabla}} \wedge \left(\underline{\tilde{\vec{E}}}_e - \frac{\partial \underline{\vec{A}}}{\partial t} \right) + \frac{\sigma_m}{\mu} (\tilde{\vec{\nabla}} \wedge \underline{\vec{A}}) = \vec{0}$$

[2.10]

In the general case of any heterogenous lossy media, the computation process reaches a blocking point. Nevertheless, in the case of uniform ratios σ/ε and σ_m/μ, it is possible to find a scalar distribution $\underline{\varphi}$ called the *electrical scalar potential* and a pseudo-scalar distribution $\underline{\varphi}_m$ called the *magnetic pseudo-scalar potential* so that

$$\underline{\tilde{\vec{H}}}_m = \frac{\sigma}{\varepsilon} \underline{\tilde{\vec{A}}}_m + \frac{\partial \underline{\tilde{\vec{A}}}_m}{\partial t} + \vec{\nabla} \underline{\tilde{\varphi}}_m \quad \text{and} \quad \underline{\tilde{\vec{E}}}_e = -\frac{\sigma_m}{\mu} \underline{\vec{A}} - \frac{\partial \underline{\vec{A}}}{\partial t} - \vec{\nabla} \underline{\varphi} \qquad [2.11]$$

The similarity between the equations of couple [2.7] and between those of couple [2.11] shows the concept of duality between electrical and magnetic potentials. This is frequently used in what follows, thus avoiding the need to repeat similar computations.

2.2.2. *Lossy propagation equations, the Lorentz gauge*

As has been required in the previous discussion, the second part of [2.8] is zero:

36 Electromagnetism and Interconnections

$$\sigma \underline{\vec{E}}_e + \varepsilon \frac{\partial \underline{\vec{E}}_e}{\partial t} - \underline{\vec{\nabla}} \wedge \underline{\vec{H}}_e = 0$$

By means of the theory of distributions, we write

$$\underline{\vec{\nabla}} \wedge \underline{\vec{H}}_e = \{\underline{\vec{\nabla}} \wedge \underline{\vec{H}}_e\} + \vec{n} \wedge \sigma_{\vec{H}} \delta_S = \{\underline{\vec{\nabla}} \wedge \underline{\vec{H}}_e\} + \vec{j}_s$$

Using the expression of $\underline{\vec{H}}_e$ arising from [2.7]:

$$\underline{\vec{H}}_e = \frac{1}{\mu} \underline{\vec{\nabla}} \wedge \underline{\vec{A}}$$

and the expression of $\vec{E}$ arising from [2.11] in the particular case of uniform σ/ε and σ_m/μ:

$$\underline{\vec{E}}_e = -\frac{\sigma_m}{\mu} \underline{\vec{A}} - \frac{\partial \underline{\vec{A}}}{\partial t} - \vec{\nabla}\varphi$$

we obtain

$$\left(\sigma + \varepsilon \frac{\partial}{\partial t}\right)\left(-\frac{\sigma_m}{\mu} \underline{\vec{A}} - \frac{\partial \underline{\vec{A}}}{\partial t} - \vec{\nabla}\varphi\right) - \underline{\vec{\nabla}} \wedge \left[\frac{1}{\mu} \underline{\vec{\nabla}} \wedge \underline{\vec{A}}\right] = 0$$

It is now possible to obtain from this last equation a classical "lossy propagation equation" in so far as we assume a new constraint without lack of generalization. This new constraint is an equation called the *Lorentz gauge*. Here we consider the Lorenz gauge suitable for obtaining a lossy propagation equation. Applying the formula

$$\begin{cases} \vec{\nabla} \wedge (\lambda \vec{a}) = (\vec{\nabla}\lambda) \wedge \vec{a} + \lambda(\vec{\nabla} \wedge \vec{a}) \\ \vec{\nabla} \wedge (\vec{\nabla} \wedge \vec{b}) = \vec{\nabla}(\vec{\nabla}\vec{b}) - \nabla^2 \vec{b} \end{cases}$$

yields

$$\underline{\vec{\nabla}} \wedge \left[\frac{1}{\mu}(\underline{\vec{\nabla}} \wedge \underline{\vec{A}})\right] = \vec{\nabla}\left(\frac{1}{\mu}\right) \wedge (\underline{\vec{\nabla}} \wedge \underline{\vec{A}}) + \frac{1}{\mu}\left[\underline{\vec{\nabla}}(\underline{\vec{\nabla}}\underline{\vec{A}}) - \nabla^2 \underline{\vec{A}}\right]$$

Setting this formula into the last equation shows a way of proposing the required Lorenz gauge (heterogenous media; σ/ε and σ_m/μ constants):

$$\vec{\nabla}(\vec{\nabla}\underline{\vec{A}}) + \mu\vec{\nabla}\left(\frac{1}{\mu}\right) \wedge (\vec{\nabla}\wedge\underline{\vec{A}}) + \mu\sigma\vec{\nabla}\underline{\varphi} + \mu\varepsilon\vec{\nabla}\frac{\partial\underline{\varphi}}{\partial t} = \vec{0} \qquad [2.12]$$

so that we obtain the following *lossy propagation equation* with the second part:

$$\nabla^2 \underline{\vec{A}} - (\mu\sigma + \varepsilon\sigma_m)\frac{\partial \underline{\vec{A}}}{\partial t} - \mu\varepsilon\frac{\partial^2 \underline{\vec{A}}}{\partial t^2} - \sigma\sigma_m \underline{\vec{A}} = \mu\underline{\vec{j}}_s\delta_S \qquad [2.13]$$

Because of the duality in electromagnetism, the substitutions

$$\underline{\vec{E}}_e \to -\underline{\vec{H}}_m, \quad \underline{\vec{A}} \to \underline{\vec{A}}_m, \quad \underline{\varphi} \to \underline{\varphi}_m, \quad \varepsilon \to \mu, \quad \mu \to \varepsilon, \quad \sigma \to \sigma_m$$

give the same lossy propagation equation for the magnetic pseudo-vector potential as for the electrical vector potential while the Lorentz gauge becomes (always in the case of uniform σ/ε and σ_m/μ):

$$\vec{\nabla}(\vec{\nabla}\underline{\vec{A}}_m) - \varepsilon\vec{\nabla}\left(\frac{1}{\varepsilon}\right) \wedge (\vec{\nabla}\wedge\underline{\vec{A}}_m) + \varepsilon\sigma_m\vec{\nabla}\underline{\varphi}_m + \mu\varepsilon\vec{\nabla}\frac{\partial\underline{\varphi}_m}{\partial t} = \mu\underline{\vec{j}}_{mS}\delta_S$$

When ε, μ, σ, and σ_m are uniform over the whole space, the Lorentz gauges become:

$$\vec{\nabla}\left[\vec{\nabla}\underline{\vec{A}} + \mu\sigma\underline{\varphi} + \mu\varepsilon\frac{\partial\underline{\varphi}}{\partial t}\right] = \vec{0} \qquad \vec{\nabla}\left[\vec{\nabla}\underline{\vec{A}}_m + \varepsilon\sigma_m\underline{\varphi}_m + \mu\varepsilon\frac{\partial\underline{\varphi}_m}{\partial t}\right] = \vec{0}$$

Therefore, the functions within brackets have to be constant: it is chosen that they are zero. So, we set

$$\vec{\nabla}\underline{\vec{A}} + \mu\sigma\underline{\varphi} + \mu\varepsilon\frac{\partial\underline{\varphi}}{\partial t} = 0 \qquad \vec{\nabla}\underline{\vec{A}}_m + \varepsilon\sigma_m\underline{\varphi}_m + \mu\varepsilon\frac{\partial\underline{\varphi}_m}{\partial t} = 0 \qquad [2.14]$$

The lossy propagation equations of $\underline{\vec{A}}$ and $\underline{\vec{A}}_m$ are the same, whether the media are homogenous or not. In the particular case of homogenous media, there are lossy propagation equations for $\underline{\varphi}$ and $\underline{\varphi}_m$.

Indeed, using

$$\vec{\nabla} \wedge (\vec{\nabla} \wedge \vec{b}) = \vec{\nabla}(\vec{\nabla}\vec{b}) - \nabla^2 \vec{b} \quad \text{and} \quad \vec{\nabla}(\vec{\nabla} \wedge \vec{c}) = 0$$

yields the equations

$$\vec{\nabla}(\nabla^2 \underline{\vec{A}}) = \vec{\nabla}\left[\vec{\nabla}(\vec{\nabla}\underline{\vec{A}}) - \vec{\nabla} \wedge (\vec{\nabla} \wedge \underline{\vec{A}})\right] = \nabla^2(\vec{\nabla}\underline{\vec{A}}) = -\mu\left(\sigma + \varepsilon\frac{\partial}{\partial t}\right)\nabla^2 \underline{\varphi}$$

$$\vec{\nabla}\left[(\mu\sigma + \varepsilon\sigma_m)\frac{\partial \underline{\vec{A}}}{\partial t} - \mu\varepsilon\frac{\partial^2 \underline{\vec{A}}}{\partial t^2} - \sigma\sigma_m \underline{\vec{A}}\right]$$

$$= -\mu\left(\sigma + \varepsilon\frac{\partial}{\partial t}\right)\left[(\mu\sigma + \varepsilon\sigma_m)\frac{\partial \underline{\varphi}}{\partial t} - \mu\varepsilon\frac{\partial^2 \underline{\varphi}}{\partial t^2} - \sigma\sigma_m \underline{\varphi}\right]$$

The application of the operator $\vec{\nabla}$ to the parts of the telegrapher's equation for $\underline{\vec{A}}$ leads to

$$-\mu\left(\sigma + \varepsilon\frac{\partial}{\partial t}\right)\left[\nabla^2 \underline{\varphi} - (\mu\sigma + \varepsilon\sigma_m)\frac{\partial \underline{\varphi}}{\partial t} - \mu\varepsilon\frac{\partial^2 \underline{\varphi}}{\partial t^2} - \sigma\sigma_m \underline{\varphi}\right] = \vec{\nabla}(\mu\vec{j}_s\delta_s) = \mu\vec{\nabla}(\vec{j}_s\delta_s)$$

Now we recall surface continuity equation [2.5]:

$$\vec{\nabla}(\vec{j}_s\delta_s) + \left(\sigma + \varepsilon\frac{\partial}{\partial t}\right)(\rho_s\delta_s) = 0$$

Then, we obtain the solution

$$\nabla^2 \underline{\varphi} - (\mu\sigma + \varepsilon\sigma_m)\frac{\partial \underline{\varphi}}{\partial t} - \mu\varepsilon\frac{\partial^2 \underline{\varphi}}{\partial t^2} - \sigma\sigma_m \underline{\varphi} - \rho_s\delta_s = e^{-(\sigma/\varepsilon)t} \qquad [2.15]$$

The second part converges quickly towards zero, which leads to a new lossy propagation equation.

The duality in electromagnetism directly leads to a similar equation for $\underline{\varphi}_m$ in homogenous media as a result of the following substitutions:

$$\underline{\varphi} \to \underline{\breve{\varphi}}_m, \quad \varepsilon \to \mu, \quad \mu \to \varepsilon, \quad \sigma \to \sigma_m, \quad \sigma_m \to \sigma$$

2.2.3. Green's kernels for harmonic electromagnetic waves in heterogenous media

The linear equations of electromagnetism can be solved by taking advantage of the concept of the kernel distribution of a linear mapping reviewed in section 1.1.5 devoted to the theory of distributions. We met the linear equation written as $A^{-1}(u) = f$ and it has been shown that if there is a linear mapping B commuting with A^{-1}, then the solution is given by $u = B(G)$, where G is an elementary solution of $A^{-1}(u) = \delta$ and is called the *Green's kernel* of this equation.

In electromagnetism, as in any other domain of physics, the features involved have a finite energy so we justify the mathematical models based on the use of Hilbert spaces fed by the scalar product of functions like f and g, denoted (f, g). Let $\{\varphi_1, \varphi_2, \ldots \varphi_n, \ldots\}$ be a Hilbertian base so that

$$(\varphi_n, \varphi_m) = 0 \quad \text{if } n \neq m$$
$$(\varphi_n, \varphi_m) = 1 \quad \text{if } n = m$$

Then, we obtain the Hilbertian expansion where we point out separately the variables (x', y', z'), hereafter denoted as x', which are involved in the scalar products and the variables (x, y, z), hereafter denoted as x, which are not involved in them:

$$u(x) = \sum_n (u(x'), \varphi_n(x')) \varphi_n(x) \quad \text{and} \quad f(x) = \sum_n (f(x'), \varphi_n(x')) \varphi_n(x)$$

Assuming x has a fixed value x_0, the linear mapping T from a function to a real number:

$$f(x') \xrightarrow{T} f(x_0) = \sum_n (f(x'), \varphi_n(x')) \varphi_n(x_0)$$

defines a distribution strictly identified with the Dirac distribution $\delta(x' - x_0)$ in respect of [1.3].

Physicists [JAC5] improperly write the following:

$$\delta(x - x_0) = \sum_n \varphi_n(x) \varphi_n(x_0) \qquad [2.16]$$

which is called a *closure relation*. This closure relation is useful for finding the Green's kernels corresponding to given constraints and therefore all the solutions of linear problems in so far as the Hilbertian base is appropriately chosen as regards the operator tied to the linear mapping A^{-1}.

Indeed, if the Hilbertian expansion is

$$G(x) = \sum_n \left(G(x'), \varphi_n(x')\right) \varphi_n(x)$$

then we obtain

$$A^{-1}[G(x)] = \sum_n \left(G(x'), \varphi_n(x')\right) A^{-1}[\varphi_n(x)]$$

Obviously, it is pertinent to require that φ_n are eigenfunctions of A^{-1}: $A^{-1}\varphi_n = \lambda_n \varphi_n$ in so far as A^{-1} meets the mathematical conditions of compactness [BRE2]. Then, we obtain the *Green's kernel* in an explicit form well suited for meeting the boundary conditions to which the eigenvalues and functions correspond:

$$G(x, x') = \sum_n \frac{1}{\lambda_n} \varphi_n(x) \varphi_n(x') \qquad [2.17]$$

which is symmetric.

This is the case presented by the propagation of harmonic waves into any heterogenous media. Indeed, the lossy propagation equation for the vector potential:

$$\nabla^2 \vec{A} - \left(\mu(x)\sigma(x) + \varepsilon(x)\sigma_m(x)\right)\frac{\partial \vec{A}}{\partial t} - \mu(x)\varepsilon(x)\frac{\partial^2 \vec{A}}{\partial t^2} - \sigma(x)\sigma_m(x)\vec{A} = \mu \vec{j}_S(x)\delta_S$$

[2.18]

where the variation of the parameters of materials over space is highlighted, is transformed for harmonic waves by setting $\vec{A}(x,t) = \vec{A}_\omega(x,t)\mathrm{e}^{j\omega t}$ that yields the so-called Helmholtz equation with second part given by (the Helmholtz equation in lossy heterogenous media)

$$\nabla^2 \vec{A}_\omega - \left[\left(\mu(x)\sigma(x) + \varepsilon(x)\sigma_m(x)\right)j\omega - \mu(x)\varepsilon(x)\omega^2 + \sigma(x)\sigma_m(x)\right]\vec{A}_\omega = \mu \vec{j}_S(x)\delta_S$$

[2.19]

The following Hilbertian expansion:

$$\mu \vec{j}_{S\omega}(x)\delta_S = \sum_n \left(\mu \vec{j}_{S\omega}(x')\delta_S, \varphi_n(x')\right)\varphi_n(x)$$

leads directly to the solution by means of formula [1.4] defining δ_S:

$$\vec{A}_\omega(x) = \sum_n \frac{1}{\lambda_n(\omega)} \left(\mu \vec{j}_{S\omega}(x') \delta_s, \varphi_n(x') \right) \varphi_n(x)$$

$$= \sum_n \frac{1}{\lambda_n(\omega)} \iiint_\Omega \left(\mu \vec{j}_{S\omega}(x') \delta_s \varphi_n(x') \right) dv \, \varphi_n(x)$$

$$= \sum_n \frac{1}{\lambda_n(\omega)} \iint_{\partial \Omega} \left[\mu \vec{j}_{S\omega}(x') \varphi_n(x') \right] ds \, \varphi_n(x)$$

$$= \iint_{\partial \Omega} \underbrace{\left[\sum_n \frac{1}{\lambda_n(\omega)} \varphi_n(x') \varphi_n(x) \right]}_{G_\omega(x,x')} \mu \vec{j}_{S\omega}(x') \, ds$$

This yields:

$$\vec{A}_\omega(x) = \iint_{\partial \Omega} G_\omega(x, x') \mu \vec{j}_{S\omega}(x') \, ds \qquad [2.20]$$

A difficulty arises from the fact that the electrical current surface density is not known over the whole surface except at the source level. Using $\mu \vec{\tilde{H}}_e = \vec{\nabla} \wedge \vec{\underline{A}}$ at a point increasingly close to the surface, the magnetic field $\vec{\tilde{H}}_e$ becomes close to $\vec{j}_{S\omega} \wedge \vec{n}$, where $\vec{n}$ is the unitary vector normal to the surface. Nevertheless, we do have to take account of the derivatives versus the coordinates. We denote $\vec{\nabla}_x$ the nabla operator in respect of x. This operator is applied to the terms of the integral expression of the vector potential. Using the vector formula

$$\vec{\nabla} \wedge (\lambda \vec{a}) = (\vec{\nabla} \lambda) \wedge \vec{a} + \lambda (\vec{\nabla} \wedge \vec{a})$$

we obtain

$$\vec{\nabla}_x \wedge \left[\mu G_\omega(x, x') \vec{j}_{S\omega}(x') \right] = \vec{\nabla}_x \left[\mu G_\omega(x, x') \right] \wedge \vec{j}_{S\omega}(x')$$

So the integral expression of the vector potential becomes

$$\mu \underbrace{\vec{j}_{S\omega}(x)}_{unknown} \wedge \vec{n} = \iint_{\partial_1 \Omega} \left[\underbrace{\vec{\nabla}_x[\mu G_\omega(x,x')]}_{known} \wedge \underbrace{\vec{j}_{S\omega}(x')}_{unknown} \right] ds + \iint_{\partial_2 \Omega} \left[\underbrace{\vec{\nabla}_x[\mu G_\omega(x,x')]}_{known} \wedge \underbrace{\vec{j}_{S\omega}(x')}_{known} \right] ds$$

which is an integral equation to be solved with regard to the unknown current surface density.

A similar equation can be written for the magnetic current surface density involving both the electrical field and its magnetic vector potential.

This work deals with numerical computation through sharing the surface into small elements where the nabla operator (gradient) of Green's function and the current surface density can be approximated by polynomials: so is the discretization process of the integral equation. Another way is to choose a finite number of test functions and calculate the scalar products of each of them with the terms of the integral equation. All these methods lead to a finite dimension matrix equation aimed at the computation of the unknown coefficients of the polynomial representing the current surface density.

This works well for problems of the scattering of bodies [HAR] or radiating antennas [MOS]. Yet, the work is devoted only to harmonic waves. It is not completed by a Fourier analysis for any time varying signal, and generates too many computations and requires too much memory storage for any complex network modeling needed for electronic design automation as regards the capability of modern processors. This leads us to search for complementary theoretical approaches for complex electrical network modeling required by up-to-date electronic system design automation.

2.3. Topology of electromagnetic interferences

2.3.1. *Introduction*

Using any one of the previously discussed numerical methods of processing partial derivatives or integral equations applied to electrical network modeling, we are led to calculations involving very large matrices consisting of elements most of which are negligible. The non-negligible elements depend on the *physically significant interferences* in respect of electrical performances required by specifications. These matrices are computed so that the simulation of the electromagnetic field takes into account the propagation of parasitic signals due to closer and closer interferences on lines progressively further away from the initially disturbing lines.

The purpose of the present discussion is to obtain the theoretical tools necessary for topological modeling of electromagnetic interferences in any complex electrical network before the numerical computation of the matrices, like those shown in Figures 5.15, 5.16, 6.5 and 9.1, involved in the propagation equations of the electromagnetic field (equations [5.26], [6.7], [7.45] and [9.24]).

2.3.2. Topological modeling of electromagnetic interferences

The topological analysis of electromagnetic interferences is based on the theory of oriented graphs. In what follows, we explain first how to apply the theory of oriented graphs to electromagnetic interferences analysis in any electronic system in order to find a network partitioning process. Applied to crosstalk computations, this work leads to matrix profiles close to diagonal matrices in the numerical processing of Maxwell's equations.

Let N be a given set of points, so-called *nodes*; an interconnection network is a subset of $(N \times N)$ which is a set of "node" couples.

When considering the interconnections between points called "nodes" in an electronic system, each *net* is defined as a conductor connecting at least two nodes at the same continuous electrical voltage while the part of a net connecting at most two nodes at the same continuous electrical voltage is called a *wire*. In any interconnection, each wire is broken into many straight lines or *segments*.

Let us begin with an essential definition: ε being any positive real number, the *physically significant electromagnetic influence area* around any part A of the space buried by conductors is defined by the domain $D(A)$ such that:

$$M \in D(A): \ \|G(M, M')\| > \varepsilon \ \forall \ M' \in A$$

where G is the Green's kernel attached to the "elementary solution" (section 2.2.3) of the telegrapher's or lossy propagation equation [2.18] and $\|G\|$ is its modulus.

Then, the *electromagnetic influence* of a net A on another one B is called *physically significant* if:

$$B \cap D(A) \neq \emptyset$$

Let us apply the theory of oriented graphs to electromagnetic interferences. E is a set in which the elements are nets (or "conductors") of a multilayered substrate. A subjective mapping V from E on to E is such that

$$A \in E \xrightarrow{V} V(A) \in E$$

if and only if there is an electromagnetic influence of at least a net A called "active" on to the net $V(A)$ called "passive".

Then we create an *oriented graph* $\{E, V\}$ of which the vertices are the nets like A and the oriented segments are physically significant electromagnetic disturbances like $\{A, V(A)\}$.

Before going on to the applications of oriented graphs, we look at the oriented graph $\{E, V^k\}$, k being an integer number $k \in |N|$, which can be defined recursively as

$$V^{k-1}(A) \in E \xrightarrow{V} V^k(A) \in E \quad \text{with} \quad V^0(A) = A, \; V^1(A) = V(A)$$

This means that the oriented graph $\{E, V^k\}$ only contains the vertices of $\{E, V\}$ connected by a sequence of k adjacent edges of $\{E, V\}$.

The set $V^0(A) \cup V^1(A) \cup V^2(A) \cup ... \cup V^k(A)$ contains all the vertices of $\{E, V\}$ connected by at the most k edges of $\{E, V\}$.

Remembering that the set $\bigcup_k V^k(A)$ is related to the *transitive enclosure* of A and taking account of only the physically significant interferences, we are led to consider truncated transitive enclosures that are limited to the level m according to the expression

$$T_m(A) = \bigcup_{k \leq m} V^k(A)$$

The mapping $A \xrightarrow{T_m} T_m(A)$ creates the oriented graph $\{E, T_m\}$. Such a graph becomes a tree if any electromagnetic interferences coming back from disturbed nets on any active net are negligible. In contrast, $\{E, T_m\}$ is *strongly connected* if any net can be disturbed by at least one other net belonging to E.

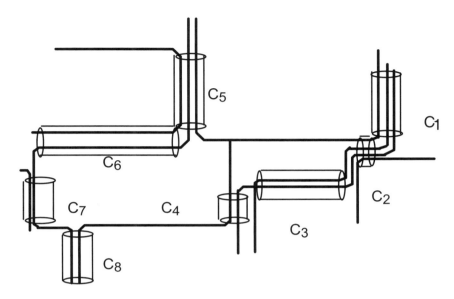

Figure 2.3. *Topology of electromagnetic interferences inside any electronic system*

2.3.3. Partitioning the electrical network in respect of electromagnetic interferences

Let us now continue to a more complete topological modeling. A set of parallel physical wires, having the same geometrical length $\{W_1, W_2,... W_r\}$ so that the graph $\left\{\bigcup_{i=1}^{i=r} W_i, T_k\right\}$ is "strongly connected", will generate a *topological cylinder* of size r related to the mapping T_k like C1 in Figure 2.3. A *topological net* is a set of "wires" connected in series without any bifurcation. So any physical net having bifurcations has to be shared into as many topological nets as bifurcations.

The union of two "topological cylinders" is a new one if and only if they are crossed by common "topological nets" whose two "topological wires" each belonging to a topological cylinder have a non-empty intersection. Figure 2.3 shows in particular a topological cylinder such as C1 which is made of two physical cylinders in series, one oriented in the X direction, the other oriented in the Y direction. In contrast, the union of topological cylinders is only a "strongly connected" graph $\{E, T_k\}$.

Then, the method of partitioning any electrical network with respect to the electromagnetic physically significant interferences is to choose first an active net, and then to find all the topological cylinders related to the mapping T_m containing this active net. Indeed the following describes the relationship between nets: "two nets having at least one wire each belonging to the same topological cylinder define an equivalence relation in the set of nets", so creating the equivalence classes giving a partition of this set. Physically, an equivalent class is the set of all the nets having mutual electromagnetic interferences with an active net that therefore does not receive any electromagnetic disturbance from any net outside the equivalent class.

The result of the simulation of electronic interferences, the graph of which is reduced to a single topological cylinder, is to get the geometrical features of the cross-section of such a cylinder depending on the physically significant electromagnetic coupling.

So a topological cylinder corresponds to a critical vicinity around many given parallel wires having the same length. To illustrate this, Figure 2.3 shows an example of complex topology of electromagnetic interferences inside an "up-to-date" electronic system.

2.3.4. *The tree of electromagnetic interferences and the problem of loops*

Since the upper mapping T_m is known thanks to a previous electrical simulation of *m* parallel lines having the same length, it is required to accurately define all the topological nets and their wires belonging to strongly connected graphs or to their cylinders.

The first step is to assign a first common rank *i* to all the nets belonging to the same strongly connected graph $\{E, T_m\}$ to which the active net belongs.

The second step is to assign a secondary common rank *j* to each net of $\{E, T_m\}$ so that any net belonging to the "topological layer" with the rank *j* defined by $V^j(A_0)$, $j \leq m$ (A_0 being the first active net), has the secondary rank *j*. As $V^0(A_0) = A_0$, the first active net has the secondary rank 0. Rank *j* can be called the *topological layer rank* in the strongly connected graph $\{E, T_m\}$.

The third step is to assign a tertiary individual rank *k* to each net belonging to the topological layer of rank *j* in the strongly connected graph $\{E, T_m\}$.

The last step is to assign a quaternary individual rank *l* to each topological cylinder of which the strongly connected graph is made. This quaternary rank has to be recursively computed. To start, note that a net is completely defined by knowledge of three integer numbers $\{i, j, k\}$ so that it can be designated as A_{ijk} or

Full Wave Analysis   47

$A(i, j, k)$. Similarly, a "topological wire", being completely defined by knowledge of four integer numbers $\{i, j, k, l\}$, can be designated as W_{ijkl} or $W(i, j, k, l)$, and

$$\forall l : W_{ijkl} \subset A_{ijk}$$

According to our theory, we say mathematically that two nets like A_{ijk} are in crosstalk as follows. $V(W_{ijkl})$ belongs to $\{E, T_m\}$, then there exists at least one pair of integer numbers (k, k')

so that: $\left. \begin{array}{l} V(A_{ijk}) \cap V(A_{i,j+1,k'}) \neq \varnothing \\ \text{for any } j : \ 0 \leq j \leq m-1 \end{array} \right\}$ [2.21]

In the same manner, two wires like W_{ijkl} are in crosstalk if there is at least one pair of integer numbers (l, l') so that

$$V(W_{ijkl}) \cap V(W_{i,j+1,k',l'}) \neq \varnothing \quad [2.22]$$

for a given $j > 0$ with $j < m$.

Let $L \times L'$ be the set of pairs $\{l \in L, l' \in L'\}$ satisfying the relationship [2.22]: each pair physically corresponds to a significant electromagnetic interference area between two nets respectively defined by the triplets (i, j, k) and $(i, j + 1, k')$. Four cases have to be considered.

Case 1. The two nets have only one wire having a significant electromagnetic interference with the other. So $L \times L'$ has only one element (l_1, l'_1), as in Figure 2.4. Then we can choose:

$$l_1 = l'_1 \quad [2.23]$$

and the ranking of each wire follows the order of meeting it by a walk on each net $A_{ijk}, A_{i,j+1,k'}$ in a suitable direction.

Case 2. The two nets have two wires in significant electromagnetic interference. So $L \times L'$ has two elements (l_1, l'_1) and (l_2, l'_2), as in Figure 2.5. Let us assume that $l_2 > l_1$, $l'_2 > l'_1$. In this case, it is always possible to draw an oriented path (for any electromagnetic pulse) along each net $A_{ijk}, A_{i,j+1,k'}$ so that ranking rule [2.23] becomes

$$\left.\begin{aligned}&l_1 = l'_1 \\ &\inf(l_2, l'_2) \to \inf(l_2, l'_2) + \sup(l_2 - l_1, l'_2 - l'_1) - \inf(l_2 - l_1, l'_2 - l'_1) \\ &\sup(l_2, l'_2) \to \sup(l_2, l'_2)\end{aligned}\right\}$$ [2.24]

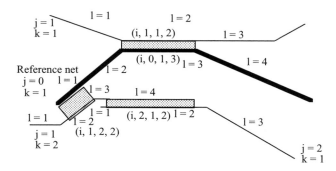

Figure 2.4. *Numbering topological cylinders depicted by shaded areas*

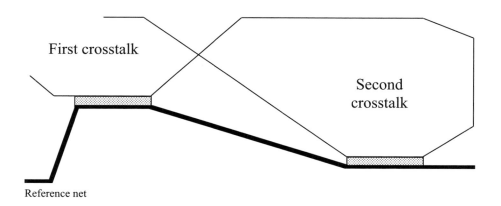

Figure 2.5. *Example of two crosstalks*

Case 3. $L \times L''$ has at least three elements (l_p, l'_p), $1 \le p \le r$, $r \ge 3$, as in Figure 2.6. The ranking rule [2.24] is successively applicable to each pair of elements $((l_p, l'_p), (l_{p+1}, l'_{p+1}))$ from $p = 1$ to $p = r - 1$, if and only if there are oriented paths on each net A_{ijk}, $A_{i,j+1,k'}$ so that

$$l_1 < l_2 < \cdots < l_p < \cdots < l_r \quad \text{and} \quad l'_1 < l'_2 < \cdots < l'_p < \cdots < l'_r$$

Then, the nets A_{ijk}, $A_{i,j+1,k'}$ are called *simultaneously orientable*.

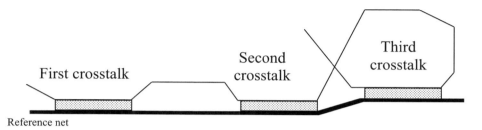

Figure 2.6. *Example of three crosstalks*

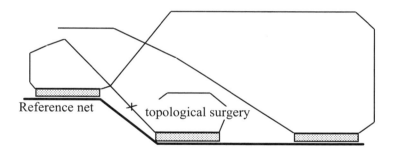

Figure 2.7. *Example of two nets that are non-simultaneously orientable*

Case 4. When the nets $A_{i,j+1,k'}$ and A_{ijk} have significant electromagnetic interferences with at least three wires each and are not simultaneously orientable, as in Figure 2.7, then we are led to do a *topological surgery* on $A_{i,j+1,k'}$ which consists of sharing it in subnets, each of them being simultaneously orientable with active net A_{ijk}. Thus, each subnet has to be considered as a bifurcation member and ranked as a new topological net. In this case, we are obliged to return to the third step of numbering the nets and their segments.

Finally, we add the following rule about k': $k' = 1$ for the first net $A_{i,j+1,k'}$ in significant crosstalk with the segment W_{ijkl} of A_{ijk} having the lowest value l_1 of l.

Then, we choose $k' = 2$ for the second net, distinct from the first one, corresponding to the lowest value l_2 of l, greater than l_1, of a segment W_{ijkl} of A_{ijk} which is not the last segment, and so on.

Some negative numbers l can occur for some segments like $W_{i,j+1,k',l'}$. Thus, a suitable numerical shift is necessary to obtain only non-negative integer numbers.

2.4. Conclusion

Electromagnetic fields propagate in relation with discontinuities arising from the boundaries between dielectric and conductors in interconnection media.

These discontinuities, where current or charges occur, behave as a break of symmetry in the space of an electromagnetic field and which involve the phenomena of propagation.

However, the rigorous integral formulations of the full wave analysis and their numerical approach, either in the time domain or frequency domain, cannot be practically compatible with the complexity of modern interconnection networks. This complexity has to be artificially reduced, by means of a topological analysis of electromagnetic interferences between interconnection network conductors, before any computation of the electromagnetic field.

Chapter 3

Electromagnetism in Stratified Media

A *stratified* electromagnetic medium is typically made of stacked layers of materials like those inside the multilayered substrates used for microelectronics. To account for the particular structure of stratified electromagnetic media, a local *electromagnetic Frenet trihedron* linked to the electrical or magnetic vector potential lines is defined. Then Maxwell's equations used in their integral and differential forms are converted into general integro-differential equations for the electrical (respectively magnetic) current flowing through any surface orthogonal to the electrical (respectively magnetic) vector potential lines that are generally bent and lying on any closed loop. This allows an advanced characterization of an extended concept of propagation modes in comparison with the classical case of straight homogenous waveguides or transmission lines.

A *full wave* analysis is developed in the particular case of straight heterogenous stratified media by means of Maxwell's equations. Then, a variational approach of the propagation equation related to the longitudinal electrical field along the axis of the media is developed and processed by analogy with the Schrödinger equation in quantum mechanics. So, a spectral analysis of all the coordinates of the electrical and magnetic fields can be done, thus yielding the required final generalized transmission line equations.

3.1. Electrical and magnetic currents in stratified media

3.1.1. *Scope of the theory, defining stratified media*

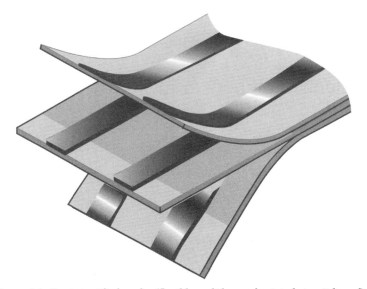

Figure 3.1. *Bent stratified media (flexible multilayered printed circuit board)*

Interconnection multilayered substrates, like printed circuit boards (Figure 3.1), behave as a stack of *bent "stratified" heterogenous media* of electromagnetic wave propagation. A heterogenous medium of electromagnetic wave propagation is said to be stratified when its characteristics of magnetic permeability, dielectric permittivity and electrical conductivity only change in each cross-section of a cylinder with straight or bent generating lines and remain uniform along the generating lines such as axis q_3 in Figure 3.2.

The following are some other concrete examples of heterogenous stratified media: waveguides limited by resistive conductors, lossy dielectric waveguides, and any multilayered interconnection substrates of which the dielectric layers have a permittivity that changes with depth from an interface (humidity absorption through coating layer, impurity diffusion inside an optical fiber, etc.).

In what follows, Maxwell's equations of electromagnetic fields inside stratified media will be transformed in order to achieve electrical signal modeling inside even *bent multilayered interconnection substrates*. The principle of our theoretical approach is based on the development of the electromagnetic field in a local trihedron attached to the electrical or magnetic vector according to the propagation mode of electromagnetic waves.

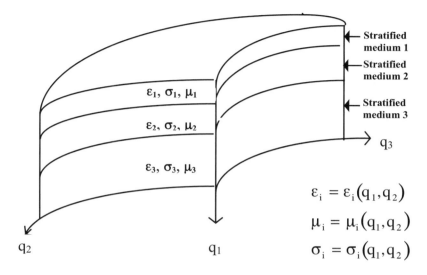

Figure 3.2. *Section of a multilayered interconnection substrate along curvilinear axes* (q_1, q_2, q_3)

Maxwell's equations are then converted into transmission line equations for any, even bent, heterogenous stratified media with any propagation mode of electromagnetic waves.

3.1.2. *Integral formulation of the current derivative versus time: general case*

Figure 3.3 illustrates the classical application of Ampere's theorem (section 1.3.7) giving the total electrical current J, including conduction or Ohm and displacement or Maxwell (section 2.1.1) currents flowing across any area set on a closed curve (C_H):

$$J = \oint_{C_H} \vec{H} \, d\vec{l} \qquad [3.1]$$

where $\vec{H}$ is the magnetic field, which is not obviously tangential to (C_H).

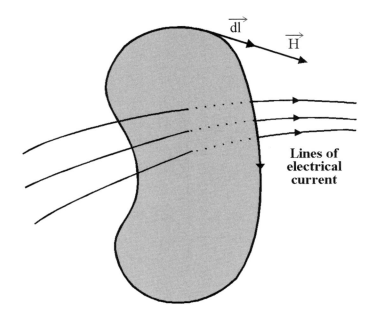

Figure 3.3. *Integration path of a magnetic field*

Differentiating J with respect to time,

$$\frac{\partial J}{\partial t} = \frac{\partial}{\partial t}\left[\oint_{C_H} \vec{H}\, \vec{dl}\right]$$

we take care with the commutability of the product of partial differential operators and curvilinear integration since (C_H) is dependent on time in the general case. If we assume that the perimeter of (C_H) is constant, then it can be demonstrated that

$$\frac{\partial J}{\partial t} = \oint_{C_H} \frac{\partial}{\partial t}(\vec{H}\, \vec{dl}) \qquad [3.2]$$

Indeed, the following is a proof of [3.2]. In the general case where the perimeter (C_H) is not constant, let us choose a trihedron based on orthogonal unitary vectors $\{\vec{u}_1, \vec{u}_2, \vec{u}_3\}$ so defining curvilinear coordinates (q_1, q_2, q_3) on which vectors like $\vec{H}$

and $\vec{dl}$ depend. These vectors also depend on time and have the following expansions:

$$\vec{H} = H_1(q_1,q_2,q_3,t)\vec{u}_1 + H_2(q_1,q_2,q_3,t)\vec{u}_2 + H_3(q_1,q_2,q_3,t)\vec{u}_3$$

$$\vec{dl} = \vec{u}_1 dq_1 + \vec{u}_2 dq_2 + \vec{u}_3 dq_3$$

Now, let us consider a curvilinear integral such as

$$J[M(t), N(t)] = \int_{M(t)}^{N(t)} \vec{H} \vec{dl}$$

along a given time-dependent rectifiable arc which is defined by the time-dependent set of functions $\{q_1(\sigma,t), q_2(\sigma,t), q_3(\sigma,t)\}$, where σ is the length measured from $M(t)$ to the running point on it. Then, the integral can be written as

$$J(t) = \int_0^{\sigma_N(t)} f(\sigma,t) \, d\sigma$$

with

$$\begin{aligned}f(\sigma,t) &= H_1[q_1(\sigma),q_2(\sigma),q_3(\sigma),t]q_1'(\sigma) + H_2[q_1(\sigma),q_2(\sigma),q_3(\sigma),t]q_2'(\sigma) \\ &+ H_3[q_1(\sigma),q_2(\sigma),q_3(\sigma),t]q_3'(\sigma)\end{aligned}$$

By means of the formula of the derivation of an integral with variable bounds, we obtain

$$J'(t) = \int_0^{\sigma_N(t)} \frac{\partial f}{\partial t} d\sigma + f(\sigma_N,t)\sigma_N'(t)$$

Formula [3.2] holds since $\sigma_N'(t) = 0$. So, $\sigma_N(t)$ is constant. If M and N are in the same location, the arc becomes a loop and $\sigma_N(t)$ is its perimeter.

Since the magnetic induction $\vec{B}$ related to $\vec{H}$ by the relationship $\vec{B} = \mu\vec{H}$ (where μ is magnetic permittivity assumed to be independent of time, but to depend on the spatial coordinates in the general case) satisfies the equation $\nabla\vec{B} = 0$, classically the vector potential $\vec{A}$ is used so that

56   Electromagnetism and Interconnections

$$\vec{H} = \frac{1}{\mu}(\vec{\nabla} \wedge \vec{A}) \qquad [3.3]$$

where ∇ (nabla) is the differential operator and $\vec{A}$ is the vector electrical potential.

The chosen method for electromagnetic field analysis is based on the expansion in the Frenet local trihedron attached to the field lines of the vector potential.

Let us define the *Frenet local trihedron* as follows (according to Figure 3.4). $\vec{n}$ is an unitary vector tangent to a field line of the vector potential $\vec{A}$. The modulus of $\vec{n}$ is equal to 1, and the modulus of $\vec{A}$ is defined by

$$\vec{A} = A\vec{n} \qquad [3.4]$$

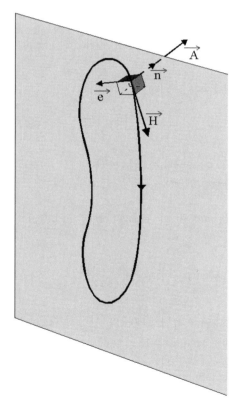

Figure 3.4. *Local electromagnetic field structure*

$\vec{e}$ is the unitary vector of the primary normal [CAG] to the field line orthogonal to $\vec{n}$.

$\vec{h}$ is the unitary vector of the secondary normal [CAG] to the field line orthogonal to both $\vec{e}$ and $\vec{n}$.

Let (C_H) be a closed curve so that

$$\mathrm{d}\vec{l} = \vec{h}\mathrm{d}l_H \qquad [3.5]$$

Let us choose

$$\vec{e} = \vec{n} \wedge \vec{h} \qquad [3.6]$$

Then, the electrical field can be written as

$$\vec{E} = E_s\vec{e} + E_n\vec{n} + E_h\vec{h} \qquad [3.7]$$

Furthermore, according to the Maxwell-Ampere equation

$$\vec{\nabla} \wedge \vec{E} = -\frac{\partial \vec{B}}{\partial t} \quad \text{and} \quad \frac{1}{\mu}(\vec{\nabla} \wedge \vec{A})$$

the electrical field $\vec{E}$ can be related to the vector potential $\vec{A}$ using the scalar potential V as follows:

$$\vec{E} = -\frac{\partial \vec{A}}{\partial t} - \vec{\nabla}V \qquad [3.8]$$

So, local coordinates are

$$E_n = -\frac{\partial A}{\partial t} - (\vec{n}\vec{\nabla})V \qquad [3.9]$$

$$E_s = -(\vec{e}\vec{\nabla})V \qquad [3.10]$$

$$E_h = -(\vec{h}\vec{\nabla})V \qquad [3.11]$$

Coming back to the derivative of the total electrical current versus time, we obtain

$$\frac{\partial J}{\partial t} = \oint_{C_H} \frac{1}{\mu} \frac{\partial}{\partial t} [\vec{h}(\vec{\nabla} \wedge \vec{A})] \, dl_H$$

The entity $\vec{h}(\vec{\nabla} \wedge \vec{A})$ can be developed into two terms by means of the formula $\vec{\nabla} \wedge (\lambda \vec{a}) = (\vec{\nabla}\lambda) \wedge \vec{a} + \lambda(\vec{\nabla} \wedge \vec{a})$:

$$\vec{h}(\vec{\nabla} \wedge \vec{A}) = \vec{h}[\vec{\nabla} \wedge (A\vec{n})] = \vec{h}[\vec{\nabla}(A) \wedge \vec{n}] + \vec{h}[A(\vec{\nabla} \wedge \vec{n})] \qquad [3.12]$$

The first term is the mixed vector product:

$$\vec{h}[\vec{\nabla}(A) \wedge \vec{n}] = (\vec{n} \wedge \vec{h})(\vec{\nabla}A) = \vec{e}(\vec{\nabla}A) = (\vec{e}\vec{\nabla})A \qquad [3.13]$$

because of [3.6]. The second term is the mixed vector product $\vec{h}(\vec{\nabla} \wedge \vec{n})$ that we have to transform as follows:

from $\vec{e} = \vec{n} \wedge \vec{h}$, we obtain $\vec{h} = \vec{e} \wedge \vec{n}$ [3.14]

According to the properties of the mixed vector product $(\vec{a},\vec{b},\vec{c}) = (\vec{a} \wedge \vec{b})\vec{c} = \vec{a}(\vec{b} \wedge \vec{c})$, we obtain

$$\vec{h}(\vec{\nabla} \wedge \vec{n}) = (\vec{e} \wedge \vec{n})(\vec{\nabla} \wedge \vec{n}) = (\vec{e},\vec{n},\vec{\nabla} \wedge \vec{n}) = \vec{e}[\vec{n} \wedge (\vec{\nabla} \wedge \vec{n})] \qquad [3.15]$$

We would like to obtain another expression for the double vector product of this last relation. In order to find it, let us recall here that

$$(\vec{n})^2 = 1 \qquad [3.16]$$

Let $\vec{a} = \vec{b} = \vec{n}$, then the vector analysis formula

$$\vec{\nabla}(\vec{a}\vec{b}) = \vec{a}(\vec{\nabla} \wedge \vec{b}) + \vec{b} \wedge (\vec{\nabla} \wedge \vec{a}) + (\vec{b}\vec{\nabla})\vec{a} + (\vec{a}\vec{\nabla})\vec{b}$$

yields the relation

$$\vec{\nabla}[(\vec{n})^2] = 2\vec{n} \wedge (\vec{\nabla} \wedge \vec{n}) + 2(\vec{n}\vec{\nabla})\vec{n} = \vec{0} \qquad [3.17]$$

because $(\vec{n})^2 = 1$. Then $\vec{n} \wedge (\vec{\nabla} \wedge \vec{n}) = -(\vec{n}\vec{\nabla})\vec{n}$ and $\vec{h}(\vec{\nabla} \wedge \vec{n}) = \vec{e}[\vec{n} \wedge (\vec{\nabla} \wedge \vec{n})]$ becomes

$$\vec{h}(\vec{\nabla} \wedge \vec{n}) = -\vec{e}[(\vec{n}\vec{\nabla})\vec{n}] \qquad [3.18]$$

Let us look at $(\vec{n}\vec{\nabla})\vec{n}$: it is the differential of the unitary vector tangent to the field line of the vector potential versus its curvilinear coordinate along this line. From classical differential geometry, it is the *curvature vector* [CAG] of this line, this vector being oriented by its primary normal vector as $\vec{e}$. Therefore

$$\vec{e}[(\vec{n}\vec{\nabla})\vec{n}] = \frac{1}{\Re} = \text{modulus of curvature} \qquad [3.19]$$

where $\Re$ is the *radius of curvature* of the vector potential line.

Recalling $\vec{h}[(\vec{\nabla}A) \wedge \vec{n}] = (\vec{e}\vec{\nabla})A$, [3.18] and [3.19] yield

$$\vec{h}(\vec{\nabla} \wedge A) = (\vec{e}\vec{\nabla})A - \frac{A}{\Re} \qquad [3.20]$$

Then, we consider the differential with respect to time:

$$\frac{\partial}{\partial t}[\vec{h}(\vec{\nabla} \wedge \vec{A})] = \frac{\partial}{\partial t}\left[(\vec{e}\vec{\nabla})\vec{A} - \frac{A}{\Re}\right] = \left[-\frac{1}{\Re} + (\vec{e}\vec{\nabla})\right]\left(\frac{\partial A}{\partial t}\right) + \left[\left(\frac{\partial}{\partial t}(\vec{e}\vec{\nabla})\right) - \frac{\partial}{\partial t}\left(\frac{1}{\Re}\right)\right]A$$

Using the local coordinate

$$E_n = -\frac{\partial A}{\partial t} - (\vec{n}\vec{\nabla})V$$

we have

$$\frac{\partial A}{\partial t} = -E_n - (\vec{n}\vec{\nabla})V$$

Note that the scalar product $(\vec{e}\vec{\nabla})(\vec{n}\vec{\nabla})$ is not commutative in the general case of curvilinear coordinates.

Indeed, in the local trihedron $\{\vec{n}, \vec{e}, \vec{h}\}$, the coordinates of the operator $\vec{\nabla}$ are such that

$$\left\{\frac{1}{e_n}\frac{\partial}{\partial q_n}, \frac{1}{e_1}\frac{\partial}{\partial q_1}, \frac{1}{e_2}\frac{\partial}{\partial q_2}\right\}$$

where e_n, e_1, e_2 are defined by the length element $ds^2 = e_1^2 dq_1^2 + e_2^2 dq_2^2 + e_n^2 dq_n^2$ depending on curvilinear coordinates q_1, q_2, q_3. We have

$$(\vec{e}\vec{\nabla})(\vec{n}\vec{\nabla}) = \frac{1}{e_1}\frac{\partial}{\partial q_1}\left(\frac{1}{e_n}\frac{\partial}{\partial q_n}\right)$$

while

$$(\vec{n}\vec{\nabla})(\vec{e}\vec{\nabla}) = \frac{1}{e_n}\frac{\partial}{\partial q_n}\left(\frac{1}{e_1}\frac{\partial}{\partial q_1}\right)$$

unequal expressions except if e_n is independent of q_1 and e_1 is independent of q_n. In this last case, both expressions become equal to

$$\frac{1}{e_1 e_n}\frac{\partial^2}{\partial q_1 \partial q_n}$$

then the commutability of $(\vec{e}\vec{\nabla})(\vec{n}\vec{\nabla})$ occurs.

Thus, we cannot carry the computations further than the following:

$$\left[-\frac{1}{\Re}+(\vec{e}\vec{\nabla})\right]\left(\frac{\partial A}{\partial t}\right) = \left[\frac{1}{\Re}-(\vec{e}\vec{\nabla})\right][E_n + (\vec{n}\vec{\nabla})V] \qquad [3.21]$$

Coming back to

$$\frac{\partial J}{\partial t} = \oint_{C_H} \frac{1}{\mu}\frac{\partial}{\partial t}[\vec{h}(\vec{\nabla}\wedge\vec{A})]\,\mathrm{d}l_H$$

results [3.20] and [3.21] yield the final required formula which is true whatever the propagation mode of electromagnetic waves inside any bent heterogenous media:

$$\frac{\partial J}{\partial t} = -\oint_{(C_H)}\frac{1}{\mu}\left[(\vec{e}\vec{\nabla})-\frac{1}{\Re}\right][E_n+(\vec{n}\vec{\nabla})V]\,\mathrm{d}l_H + \oint_{(C_H)}\frac{1}{\mu}\frac{\partial}{\partial t}\left[(\vec{e}\vec{\nabla})-\frac{1}{\Re}\right]A\,\mathrm{d}l_H \qquad [3.22]$$

The following is the physical meaning of [3.22]. The first integral is always non-zero for any mode of propagation whether or not the direction of the vector potential depends on time. This is the case of *TM* ($H_n = 0$) or *TEM* ($H_n = E_n = 0$) *propagation modes* around *weakly bent conductors* in homogenous media.

Electromagnetism in Stratified Media 61

The second integral is not zero for the modes for which the direction of the vector potential depends on time. This is the case of *TE propagation modes ($E_n = 0$)* around weakly bent conductors in homogenous media.

3.1.3. *Integral formula of the current derivative relative to space in the direction of the vector potential*

Returning to Ampere's law [3.1], the differentiation relative to space in the direction $\vec{n}$ of the vector potential $\vec{A}$ is done by means of the application of the scalar operator $(\vec{n}\vec{\nabla})$:

$$(\vec{n}\vec{\nabla})J = (\vec{n}\vec{\nabla})\oint_{C_H} \vec{H} \, d\vec{l} \qquad [3.23]$$

Assuming that the closed curve (C_H) is defined as not depending on the curvilinear coordinate in the direction $\vec{n}$, we can write the following commutability relationship between operators:

$$(\vec{n}\vec{\nabla})\oint_{C_H} \vec{H} \, d\vec{l} = \oint_{C_H} (\vec{n}\vec{\nabla})(\vec{H} \, d\vec{l}) \qquad [3.24]$$

Taking into account the definition of dl_H and the formula $(\vec{a}\vec{\nabla})(\vec{b}\vec{\nabla}) = \vec{c}(\vec{a}\vec{\nabla})\vec{b} + \vec{b}(\vec{a}\vec{\nabla})\vec{c}$, [3.24] becomes

$$(\vec{n}\vec{\nabla})J = \oint_{C_H} (\vec{n}\vec{\nabla})(\vec{H}\vec{h}) \, dl_H \qquad [3.25]$$

$$= \oint_{C_H} \vec{h}(\vec{n}\vec{\nabla})\vec{H} \, dl_H + \oint_{C_H} \vec{H}(\vec{n}\vec{\nabla})\vec{h} \, dl_H \qquad [3.26]$$

The second integral in [3.26] is zero because $h^2 = 1$. Indeed, the formula $(\vec{a}\vec{\nabla})(\vec{b}\vec{c}) = \vec{c}(\vec{a}\vec{\nabla})\vec{b} + \vec{b}(\vec{a}\vec{\nabla})\vec{c}$ gives $(\vec{n}\vec{\nabla})(\vec{h})^2 = 2\vec{h}(\vec{n}\vec{\nabla})\vec{h}$. Then,

$$(\vec{h})^2 = 1 \Rightarrow (\vec{n}\vec{\nabla})(\vec{h})^2 = 0 = 2\vec{h}(\vec{n}\vec{\nabla})\vec{h} \Rightarrow \vec{H}\vec{h}(\vec{n}\vec{\nabla})\vec{h} = \vec{H}(\vec{n}\vec{\nabla})\vec{h} = 0$$

For the first integral, we apply the formula $(\vec{a} \wedge \vec{b})(\vec{c} \wedge \vec{d}) = \vec{a}(\vec{b}\vec{c})\vec{d} - \vec{b}(\vec{a}\vec{c})\vec{d}$ in this form that allows us to replace $\vec{c}$ by $\vec{\nabla}$ with respect to the rules of the "nabla" operator algebra. Setting $\vec{a} = \vec{n}$, $\vec{b} = \vec{h}$, $\vec{d} = \vec{H}$, we obtain

62   Electromagnetism and Interconnections

$$(\vec{n} \wedge \vec{h})(\vec{\nabla} \wedge \vec{h}) = \vec{h}(\vec{n}\vec{\nabla})\vec{H} - \vec{n}(\vec{h}\vec{\nabla})\vec{H} \qquad [3.27]$$

where the vector $\vec{e}(\vec{n} \wedge \vec{h})$ is highlighted in the first term. Using this and applying the first Maxwell equation $\vec{\nabla} \wedge \vec{H} = \vec{j}$ in the first term leads to the transformation of the only remaining first integral of [3.26], thus giving the following result:

$$(\vec{n}\vec{\nabla})J = \oint_{C_H} \vec{e}\vec{j}\, dl_H + \oint_{C_H} \vec{n}(\vec{h}\vec{\nabla})\vec{H}\, dl_H \qquad [3.28]$$

Furthermore, applying the formula $(\vec{a}\vec{\nabla})(\vec{b}\vec{c}) = \vec{c}(\vec{a}\vec{\nabla})\vec{b} + \vec{b}(\vec{a}\vec{\nabla})\vec{c}$ yields

$$(\vec{h}\vec{\nabla})(\vec{n}\vec{H}) = \vec{n}(\vec{h}\vec{\nabla})\vec{H} + \vec{H}(\vec{h}\vec{\nabla})\vec{n} \qquad [3.29]$$

which leads us to write

$$\oint_{C_H} \vec{n}(\vec{h}\vec{\nabla})\vec{H}\, dl_H + \oint_{C_H} \vec{H}(\vec{h}\vec{\nabla})\vec{n}\, dl_H = \oint_{C_H} (\vec{h}\vec{\nabla})(\vec{n}\vec{H})\, dl_H + \oint_{C_H} \vec{\nabla}(\vec{n}\vec{H})\, d\vec{l} = 0$$

thus yielding the result

$$(\vec{n}\vec{\nabla})J = \oint_{C_H} \vec{e}\vec{j}\, dl_H - \oint_{C_H} \vec{H}(\vec{h}\vec{\nabla})\vec{n}\, dl_H \qquad [3.30]$$

Let us now look at $(\vec{h}\vec{\nabla})\vec{n}$:

– $\vec{h}$ is the unitary vector tangent to (C_H);

– $\vec{e}$ can be considered as the unitary vector to the primary normal of (C_H) orthogonal to the vector potential;

– $\vec{n}$ is the unitary vector of the secondary normal of (C_H).

So, from differential geometry, the derivative of a second normal relative to its curvilinear coordinate along a curve defines the *twisting vector* [CAG] of this curve in the direction of its primary normal. Following the classical approach, we can write

$$(\vec{h}\vec{\nabla})\vec{n} = -\frac{\vec{e}}{\Re_{t_H}}$$

where $\Re_{t_H}$ is the *radius of twisting* [CAG] which is finite in the general case of a skew curve (C_H).

Thus, we obtain

$$(\vec{n}\vec{\nabla})J = \oint_{C_H} \vec{e}\vec{j}\, dl_H + \oint_{C_H} \frac{\vec{e}\vec{H}}{\Re_H}\, dl_H \qquad [3.31]$$

Knowing the basic relationship related to both the Ohm and Maxwell laws:

$$\vec{j} = \left(\sigma + \varepsilon \frac{\partial}{\partial t}\right)\vec{E} \qquad [3.32]$$

we obtain the final result:

$$(\vec{n}\vec{\nabla})J = \oint_{C_H} \vec{e}\left(\sigma + \varepsilon \frac{\partial}{\partial t}\right)\vec{E}\, dl_H + \oint_{C_H} \frac{\vec{e}\vec{H}}{\Re_{t_H}}\, dl_H \qquad [3.33]$$

In the most general case, we have to take care with the fact that a trihedron defined by the direction of the vector potential depends on time. This is the reason why we prefer to write [3.33] as follows, where $E_s = \vec{e}\vec{E}$:

$$(\vec{n}\vec{\nabla})J = \oint_{C_H}\left(\sigma + \varepsilon \frac{\partial}{\partial t}\right)E_s\, dl_H - \oint_{C_H} \varepsilon \vec{E}\frac{\partial \vec{e}}{\partial t}\, dl_H + \oint_{C_H} \frac{\vec{e}\vec{H}}{\Re_{t_H}}\, dl_H \qquad [3.34]$$

The following is the physical meaning of [3.34]. The first integral is always non-zero for any mode of propagation whether or not the direction of the vector potential depends on time. This is the case of *TM ($H_n = 0$) or TEM ($H_n = E_n = 0$) propagation modes* around weakly bent conductors in homogenous media.

The second integral is non-zero since this direction depends on time. This is the case of TE modes around *weakly bent conductors* in homogenous media. The third integral vanishes if (C_H) is a plane curve. This is the case in the vicinity of straight conductors.

3.1.4. *Duality between electrical and magnetic currents in lossless media*

In free space where there are no electrical charges, the Poisson equation having its second term zero shows that the electrical induction $\vec{D}$ is the curl vector of a "magnetic potential" $\vec{A}_m$. Thus, a situation is created symmetric to that discussed above.

Indeed, since

$$\vec{\nabla}\vec{D} = 0 \quad \text{with} \quad \vec{D} = \varepsilon\vec{E}$$

there exists $\vec{A}_m$ such that

$$\vec{E} = \frac{1}{\varepsilon}(\vec{\nabla} \wedge \vec{A}_m) \qquad [3.35]$$

Such a relation is in duality with

$$\vec{H} = \frac{1}{\mu}(\vec{\nabla} \wedge \vec{A})$$

In the same way, the Maxwell-Faraday equation with lack of losses

$$\vec{\nabla} \wedge \vec{H} = \vec{j} = \frac{\partial \vec{D}}{\partial t}$$

leads us to introduce the scalar magnetic potential V_m:

$$\vec{H} = \frac{\partial \vec{A}_m}{\partial t} + \vec{\nabla} V_m \qquad [3.36]$$

This relationship is in duality with

$$\vec{E} = \frac{\partial \vec{A}}{\partial t} - \vec{\nabla} V$$

in spite of a sign change.

As a consequence, it is possible to obtain directly the equations related to the choice of a local trihedron defined by the direction of the magnetic vector potential. For this purpose, it is only necessary to consider the magnetic current J_m flowing across any loop (C_E) orthogonal to $\vec{A}_m$.

So, Faraday's law can be written in the same manner as Ampere's law:

$$J_m = \oint_{C_E} \vec{E} \, d\vec{l} \qquad [3.37]$$

which is in duality with

$$J = \oint_{C_H} \vec{H}\,\mathrm{d}\vec{l}$$

How can these built-in duality relationships be helpful for electromagnetism analysis? In fact, we have seen the difficulties introduced by integration loops depending on time or on space.

In spite of using loops having a constant perimeter, the most general case, equations [3.22] and [3.34], has integrals that contain $\partial \vec{e}/\partial t$ related to a moving local trihedron.

These integrals vanish in modes of propagation where the local trihedron involved is fixed.

The interest in duality is to obtain the advantage of fixed trihedrons for any propagation mode of electromagnetic waves. Such a trihedron can consist of $\{\vec{n}_m, \vec{e}_m, \vec{h}_m\}$, where $\vec{n}_m$ is the unitary vector in the direction of $\vec{A}_m$ in the case where the direction of $\vec{A}_m$ is dependent on time.

The built-in duality relations arising from [3.22] and [3.34] are obtained by exchanging:

– ε and μ, and μ and ε
– $\vec{H}$ and $\vec{E}$, and $\vec{E}$ and $-\vec{H}$
– $\vec{A}$ and $\vec{A}_m$
– V and V_m
– J and J_m
– $\vec{n}$ and $\vec{n}_m$
– $\vec{e}$ and $\vec{h}_m$
– σ and 0
– $\Re$ and $\Re_m$
– E_n and $-H_{mn}$
– E_s and $-H_{ms}$
– $\Re_{t_H}$ and $\Re_{t_{mE}}$
– C_H and C_E.

Thus, we obtain directly

$$\frac{\partial J_m}{\partial t} = \oint_{C_E} \frac{1}{\varepsilon}\left[(\vec{e}_m \vec{\nabla}) - \frac{1}{\Re_m}\right][H_{mn} - (\vec{n}_m \vec{\nabla})V_m] \, dl_E + \oint_{C_E} \frac{1}{\varepsilon}\frac{\partial}{\partial t}\left[(\vec{e}_m \vec{\nabla}) - \frac{1}{\Re_m}\right]A_m \, dl_E$$

[3.38]

$$(\vec{n}_m \vec{\nabla})J_m = -\oint_{C_E} \mu \frac{\partial}{\partial t} H_{ms} \, dl_E + \oint_{C_E} \mu \vec{H} \frac{\partial \vec{h}_m}{\partial t} \, dl_E + \oint_{C_E} \frac{\vec{h}_m \vec{E}}{\Re_{t_{mE}}} \, dl_E \qquad [3.39]$$

Here, we note that the first integrals of [3.38] and [3.39] remain non-zero whatever the propagation mode, while the final integrals in both [3.38] and [3.39] vanish when the longitudinal magnetic field is zero (TE mode).

To conclude our discussion about duality, in the case of weakly bent homogenous media only, the electrical current formulation will be chosen in TM or TEM modes, and magnetic current formulation in TE mode.

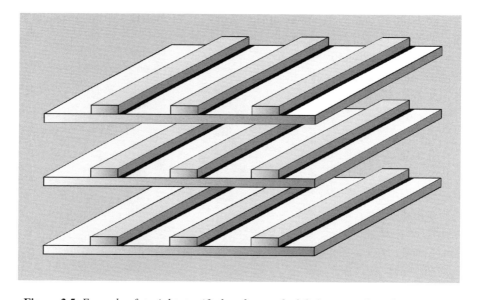

Figure 3.5. *Example of straight stratified media: stacked dielectric and conductive layers*

3.2. Straight stratified media

3.2.1. *Scope*

We are interested in straight stratified media where the dielectric permittivity, electrical conductivity, and magnetic permittivity do not vary along a straight axis but can be non-uniform in any plane orthogonal to this straight axis. Regarding the previous topological approach of electromagnetic interferences (section 2.3), any straight medium corresponds to a "topological cylinder" (section 2.3.3) containing all the physically "significant electromagnetic" perturbations (section 2.3.2) coming from a given conductor "net".

3.2.2. *Lossy propagation equations and the variational approach*

We are now interested in the case of *heterogenous stratified media* consisting of materials having different and non-uniform dielectric permittivity and electrical conductivity but the same uniform magnetic permeability. Then, we look at theoretical modeling of the electrical behavior of stacked transmission lines (Figure 3.5) where both conductors and dielectric losses are accounted for.

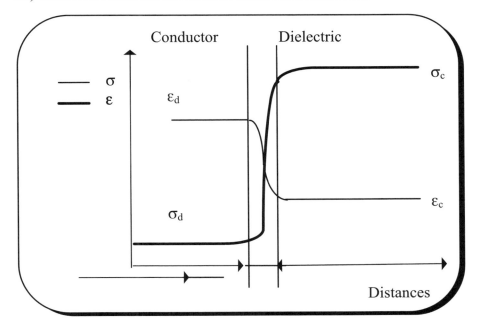

Figure 3.6. *Permittivity and conductivity variations across a dielectric–conductor interface*

For this purpose, we consider that the electrical permittivity, but more so the electrical conductivity, have rough variations at the conductor–dielectric interfaces (Figure 3.6) that are modeled by many boundary conditions coming from Maxwell's equations written according to the theory of distributions and devoted to heterogenous media as discussed in Chapter 2. In contrast to that, we assume the electrical parameters very continuously, even at the conductor–dielectric interface where the variation has a large slope.

Let us find the *lossy propagation equations* arising from Maxwell's equations. Prior to doing this, we have to recall briefly how the propagation equation is obtained for the electrical field:

$$\vec{\nabla} \wedge (\vec{\nabla} \wedge \vec{E}) = \vec{\nabla} \wedge \left(-\mu \frac{\partial \vec{H}}{\partial t}\right) = -\mu \frac{\partial}{\partial t}(\vec{\nabla} \wedge \vec{H}) - \mu\sigma \frac{\partial \vec{E}}{\partial t} - \mu\varepsilon \frac{\partial^2 \vec{E}}{\partial t^2}$$

which is still valid in any medium having heterogenous dielectric permittivity and electrical conductivity not depending on time.

Knowing that the electrical charge density is strictly zero in any homogenous medium, we have $\vec{\nabla}\vec{E} = 0$. Then $\vec{\nabla} \wedge (\vec{\nabla} \wedge \vec{E}) = \vec{\nabla}(\vec{\nabla}\vec{E}) - \nabla^2 \vec{E} = -\nabla^2 \vec{E}$, and the propagation equation dedicated to the electrical field can be written as

$$\nabla^2 \vec{E} - \mu\sigma \frac{\partial \vec{E}}{\partial t} - \mu\varepsilon \frac{\partial^2 \vec{E}}{\partial t^2} = \vec{0} \qquad [3.40]$$

This equation is not valid for the magnetic field in this case.

Indeed, in this case we have

$$\vec{\nabla} \wedge (\vec{\nabla} \wedge \vec{H}) = \vec{\nabla} \wedge \left(\sigma \vec{E} + \varepsilon\mu \frac{\partial \vec{E}}{\partial t}\right) = \left(\sigma + \varepsilon \frac{\partial}{\partial t}\right)(\vec{\nabla} \wedge \vec{E}) + \vec{\nabla}\left(\sigma + \varepsilon \frac{\partial}{\partial t}\right) \wedge \vec{E}$$

Then, in spite of Maxwell's equations, $\vec{\nabla} \wedge \vec{E} = -\mu(\partial \vec{H} / \partial t)$, the remaining term

$$\vec{\nabla}\left(\sigma + \varepsilon \frac{\partial}{\partial t}\right) \wedge \vec{E}$$

cannot meet the lossy propagation equation except in the case of homogenous media (Chapter 4).

The duality in electromagnetism leads directly to the same equation as above for the longitudinal magnetic field when the magnetic permeability is non-uniform while the dielectric permittivity and electrical conductivity are not dependent on time and are uniform.

Let us return to the case of non-uniform dielectric permittivity and electrical conductivity not depending on time with uniform magnetic permeability. In this case, accounting for the electromagnetic wave propagation along the z axis, we consider now the lossy propagation equation for the coordinate $E_z = E_n$ of the electrical field written as follows:

$$\nabla_s^2 E_n + \frac{\partial^2 E_n}{\partial z^2} = \mu\varepsilon \frac{\partial^2 E_n}{\partial t^2} + \mu\sigma \frac{\partial E_n}{\partial t}$$

where

$$\nabla_s^2 = \frac{\partial^2}{\partial x^2} + \frac{\partial^2}{\partial y^2}$$

is the Laplacian operator in the plane (x, y).

For the sake of mathematical efficiency, we apply the well-known results arising from functional analysis and focused on the variational methods of solving partial differential equations, completed by spectral expansions by with the help of the finite element approximation of eigenvalue computations.

Here, we consider that the stratified media are generally not bounded by a closed lossless conductive surface. So, lossy media modeling requires that the electromagnetic field stretches over the whole space. Nevertheless, it is known in functional analysis [DAU1] that the Laplacian operator (here ∇_s^2) involved in the lossy propagation equation has a continuous spectrum $[0; +\infty[$ in the set of square integrable functions defined inside the whole space.

We hope now to be able, using a discrete spectrum, to more easily handle the mathematics.

It is known in quantum mechanics [LAN2] that the Hamiltonian operator H defined by $-\nabla^2 + \varphi$ (where φ is an operator of a simple multiplication by a function φ strongly decreasing and also all its derivatives) has a discrete spectrum [DAU2]. This is interesting for our stratified media modeling when σ and ε and so the parameter $a = \mu\sigma^2/4\varepsilon$ are varying over a limited distance according the following law which is not necessarily periodic:

When going through the conductive and dielectric layers alternately, this shows a behavior similar to that encountered in a finite dimension crystal lattice [VAP1]. An application to Foucault's current is developed further at the end of Chapter 4. In the harmonic state, analytical computations can be carried out similar to those concerning surface waves along dielectric interfaces [ROU].

This introduction leads us to process the following transformation of the second term of the lossy propagation equation:

$$\nabla_s^2 E_n + \frac{\partial^2 E_n}{\partial z^2} = \mu\varepsilon \frac{\partial^2 E_n}{\partial t^2} + \mu\sigma \frac{\partial E_n}{\partial t} = \mu\varepsilon \left(\frac{\partial}{\partial t} + \frac{\sigma}{2\varepsilon}\right)^2 E_n - \frac{\mu\sigma}{4\varepsilon} E_n$$

Then we consider the partial differential equation suitable for applying the variational method:

$$\frac{\partial}{\partial x}\left(\frac{\partial E_n}{\partial x}\right) + \frac{\partial}{\partial y}\left(\frac{\partial E_n}{\partial y}\right) + \frac{\mu\sigma^2}{4\varepsilon} E_n + \frac{\partial^2 E_n}{\partial z^2} - \mu\varepsilon\left(\frac{\partial}{\partial t} + \frac{\sigma}{2\varepsilon}\right)^2 E_n = 0$$

Indeed, let us recall the *variational method* for solving the partial differential equation above. According to the classical process developed in functional analysis, we consider the following:

– First, we consider a test function F belonging to the "Sobolev space H^1" (which means the set of all the functions having a square integral and their derivatives as well).

– Then, multiply the terms of the partial differential equation by F.

– Integrate these terms into the whole space, then continue to the equation

$$\iint_{R^2}\left[\frac{\partial}{\partial x}\left(\frac{\partial E_n}{\partial x}\right) + \frac{\partial}{\partial y}\left(\frac{\partial E_n}{\partial y}\right)\right] F\, dxdy + \iint_{R^2} \frac{\mu\sigma^2}{4\varepsilon} E_n F\, dxdy + \iint_{R^2} \frac{\partial^2}{\partial z^2} F\, dxdy$$

$$- \iint_{R^2} \mu\varepsilon \left(\frac{\partial}{\partial t} + \frac{\sigma}{2\varepsilon}\right)^2 E_n F\, dxdy = 0$$

– Apply the classical Green's formula:

$$\iint_{R^2}\left[\frac{\partial}{\partial x}\left(\frac{\partial E_n}{\partial x}\right)+\frac{\partial}{\partial y}\left(\frac{\partial E_n}{\partial y}\right)\right]F\,dxdy = \iint_{R^2}\left(\frac{\partial E_n}{\partial x}\frac{\partial F}{\partial x}+\frac{\partial E_n}{\partial y}\frac{\partial F}{\partial y}\right)dxdy$$

– Define the *bilinear mapping* F from the set of couples $\{E_n, F\}$ to the set of real numbers:

$$F(E_n, F) = \iint_{R^2}\left(\frac{\partial E_n}{\partial x}\frac{\partial F}{\partial x}+\frac{\partial E_n}{\partial y}\frac{\partial F}{\partial y}-aE_n F\right)dxdy$$

where $a = \mu\sigma^2/4\varepsilon$ is a bounded not necessarily continuous function.

– Finally, use the notation of the *scalar product*:

$$(f, g) = \iint_{R^2} fg\,dxdy$$

Therefore, we obtain the classical equation of the variational approach:

$$\left(\mu\varepsilon\left(\frac{\partial}{\partial t}+\frac{\sigma}{2\varepsilon}\right)^2 E_n, F\right)-\frac{\partial^2}{\partial z^2}(E_n, F)+F(E_n, F) = 0$$

which is the variational form of the lossy propagation equation in any heterogenous media suited for numerical computations based on the spectral analysis that we now need to develop.

3.2.3. *Spectral analysis of the longitudinal field*

We introduce the *Hamiltonian operator* $H = H = -\nabla_s^2 - \mu\sigma^2/4\varepsilon$ which has a discrete spectrum made of eigenvalues denoted as β_j^2 in so far as the function $a = \mu\sigma^2/4\varepsilon$ is strongly decreasing and all its derivatives as well.

Yet, the eigenfunction E_n^1 which is related to the eigenvalue β_j^2 of the Hamiltonian operator is a solution of the partial differential equation:

$$HE_n^j = \beta_j^2 E_n^j \qquad [3.41]$$

Let us use the previous variational procedure as follows:

– Introduce a test function F belonging to the "Sobolev space H^1" (which means the set of all the functions having a square integral and their derivatives as well).
– Then, multiply the terms of the partial differential equation by F.
– Integrate these terms into the whole space, then continue to the equation:

$$\iint_{R^2} HE_n^j F\,dxdy = \beta_j^2 \iint_{R^2} E_n^j F\,dxdy$$

– Apply the classical Green's formula written as

$$(HE_n^j, F) = F(E_n^j, F)$$

so that the eigenfunction equation becomes

$$F(E_n^j, F) = \beta_j^2 (E_n^j, F) \qquad [3.42]$$

According to this, the electrical field can be expanded in the Hilbertian base of the above-mentioned elliptic operator as a function of (x, y):

$$E_z = E_n = \sum_j e_j(z,t) E_n^j(x, y)$$

In this case, the variational equation

$$\left(\left[\mu\varepsilon \left(\frac{\partial}{\partial t} + \frac{\sigma}{2\varepsilon} \right)^2 - \frac{\partial^2}{\partial z^2} \right] E_n, F \right) + F(E_n, F) = 0$$

becomes

$$\left(\left[\mu\varepsilon \left(\frac{\partial}{\partial t} + \frac{\sigma}{2\varepsilon} \right)^2 - \frac{\partial^2}{\partial z^2} \right] \sum_j e_j(z,t) E_n^j(x,y), F \right) + F\left(\sum_j e_j(z,t) E_n^j(x,y), F \right)$$

$$= \sum_j \left[\left(\left[\mu\varepsilon \left(\frac{\partial}{\partial t} + \frac{\sigma}{2\varepsilon} \right)^2 - \frac{\partial^2}{\partial z^2} \right] e_j E_n^j, F \right) + e_j F(E_n^j, F) \right] = 0$$

accounting for the variational equation $F(E_n^j, F) = \beta_j^2 (E_n^j, F)$.

Equation [3.41] or its similar form [3.42] is suitable for the nodal analysis of Foucault's currents discussed in section 4.3.1. Moreover, computation of the lowest eigenvalue can be done by means of the method shown in section 4.2.3. Let λ_1 be this lowest eigenvalue; applying the method shown in section 4.2.3 to the matrix denoted $[R]^{-1}[M] - \lambda_1[1]$, $[R]$ and $[M]$ being defined there, leads to the lowest eigenvalue λ_2 greater than λ_1, and so on.

So, we obtain

$$\sum_j \left[\left(\left[\mu\varepsilon\left(\frac{\partial}{\partial t}+\frac{\sigma}{2\varepsilon}\right)^2 - \frac{\partial^2}{\partial z^2}\right]e_j E_n^j, F\right) + e_j \beta_j^2 (E_n^j, F)\right] = 0$$

We choose the function F among the eigenfunctions themselves, thus accounting for the scalar product (E_n^j, E_n^i) being zero when i and j are different and equal to 1 when i and j are equal.

Let us develop the operator:

$$\mu\varepsilon\left(\frac{\partial}{\partial t}+\frac{\sigma}{2\varepsilon}\right)^2 = \mu\varepsilon\frac{\partial^2}{\partial t^2} + \mu\sigma\frac{\partial}{\partial t} + a$$

Assuming the functions $\mu\varepsilon(x,y)E_n^j(x,y)$, $(\mu/2)\sigma(x,y)E_n^j(x,y)$ and $a(x,y)E_n^j(x,y)$ belong to the Hilbert space spanned by the eigenfunction $E_n^j(x,y)$, we can write the following expansions:

$$\mu\varepsilon(x,y)E_n^j(x,y) = \sum_i \varepsilon_i^j E_n^i(x,y)$$

$$\frac{\mu}{2}\sigma(x,y)E_n^j(x,y) = \sum_i \sigma_i^j E_n^i(x,y)$$

$$a(x,y)E_n^j(x,y) = \sum_i a_i^j E_n^i(x,y)$$

where ε_i^j, σ_i^j, and a_i^j are known constant coefficients defined only by the known eigenfunctions and the known functions ε, σ and a by means of the scalar products:

$$\varepsilon_i^j = (\mu\varepsilon E_n^j, E_n^i) \tag{3.43}$$

$$\sigma_i^j = \left(\frac{\mu\sigma}{2}E_n^j, E_n^i\right) \tag{3.44}$$

$$a_i^j = (aE_n^j, E_n^i) \tag{3.45}$$

The matrices $[a]$, $[\sigma]$ and $[\varepsilon]$, having the elements a_i^j, σ_i^j and ε_i^j, respectively, are real and commutable because their elements are defined by commutative scalar products as follows: $(\mu\varepsilon E_n^j, E_n^i) = (\mu\varepsilon E_n^{ij}, E_n^j)$; so, $\varepsilon_i^j = \varepsilon_j^i$.

Moreover, they are commutable. Indeed, applying twice the formula defining their elements gives

$$\frac{\mu\sigma}{2}(\mu\sigma E_n^j) = \sum_i \varepsilon_i^j \left(\frac{\mu\sigma}{2} E_n^i\right) = \sum_i \varepsilon_i^j \sum_k \sigma_k^i E_n^k = \sum_k \left(\sum_i \varepsilon_i^j \sigma_k^i\right) E_n^k$$

$$\mu\varepsilon\left(\frac{\mu\sigma}{2} E_n^j\right) = \sum_i \sigma_i^j (\mu\varepsilon E_n^i) = \sum_i \sigma_i^j \sum_k \varepsilon_k^i E_n^k = \sum_k \left(\sum_i \sigma_i^j \varepsilon_k^i\right) E_n^k$$

The algebraic equality

$$\frac{\mu\sigma}{2}(\mu\varepsilon E_n^j) = \mu\varepsilon\left(\frac{\mu\sigma}{2} E_n^j\right)$$

leads to the matrix relation $[\sigma][\varepsilon] = [\varepsilon][\sigma]$ and also to the commutability of $[\sigma]$ and $[a]$, $[a]$ and $[\varepsilon]$.

Furthermore, they are positive definite. Indeed, the quadratic form

$$\sum_{i,j} \varepsilon_i^j E_n^j E_n^i = \sum_j \left(\sum_i \varepsilon_i^j E_n^i\right) E_n^i = \sum_i (\mu\varepsilon E_n^i) E_n^i = \mu\varepsilon \sum_i (E_n^i)^2$$

is positive, which shows the infinite dimension square matrix $[\varepsilon]$ is positive definite. So are the other infinite dimension matrices $[\sigma]$ and $[a]$ having the elements σ_i^j and a_i^j, respectively.

To conclude, we have the relation

$$[\sigma]^2 = [a][\varepsilon] = [\varepsilon][a] \qquad [3.46]$$

Indeed,

$$\left(\frac{\mu\sigma}{2}\right)^2 E_n^j = \frac{\mu\sigma}{2}\left(\frac{\mu\sigma}{2}E_n^j\right) = \frac{\mu\sigma}{2}\sum_i \sigma_i^j\left(\frac{\mu\sigma}{2}E_n^i\right) = \sum_i \sigma_i^j \sum_k \sigma_k^i E_n^k$$
$$= \sum_i \left(\sum_i \sigma_i^j \sigma_k^i\right) E_n^k$$

$$\left(\frac{\mu\sigma}{2}\right)^2 E_n^j = \frac{\mu\sigma^2}{4}(\mu\sigma E_n^j) = a\sum_i \varepsilon_i^j E_n^i = \sum_{i,j} \varepsilon_i^j (aE_n^i) = \sum_i \varepsilon_i^j \sum_k a_k^i E_n^k$$
$$= \sum_i \left(\sum_i \varepsilon_i^j a_k^i\right) E_n^k$$

from which

$$\sum_i \sigma_i^j \sigma_k^i = \sum_i \varepsilon_i^j a_k^i$$

leading to relation [3.46].

Then the lossy propagation equation becomes a generalized telegrapher's equation in the plane (z, t):

$$\frac{\partial^2 e_j}{\partial z^2} - \sum_i \left(\varepsilon_i^j \frac{\partial^2}{\partial t^2} + 2\sigma_i^j \frac{\partial}{\partial t} + a_i^j - \delta_i^j \beta_j^2\right) e_j = 0$$

where δ_i^j is the Kronecker symbol, equal to 1 if $i = j$ and zero if $i \neq j$.

This last equation can be written in matrix form:

$$\left\{\frac{\partial^2}{\partial z^2} - [\varepsilon]\frac{\partial^2}{\partial t^2} - 2[\sigma]\frac{\partial}{\partial t} - [a] + [\beta]^2\right\}[e] = 0$$

The matrix $[\beta]^2$, the elements of which are $\delta_i^j \beta_j^2$, is diagonal.

Since the matrices $[a]$ and $[\varepsilon]$ are real, symmetric and positive, they have positive eigenvalues, and it is possible to define the matrices $[a]^{1/2}$ and $[\varepsilon]^{1/2}$ according to the spectral theory of matrices [LAS1]. So, these matrices also being commutable, the previous equations become

$$\left\{ \frac{\partial^2}{\partial z^2} - \left([\varepsilon]^{1/2} \frac{\partial}{\partial t} + [a]^{1/2} \right)^2 + [\beta]^2 \right\} [e] = 0 \qquad [3.47]$$

3.2.4. *From Maxwell's equations to transmission line equations*

Now, in the general case, we need Maxwell's equations [1.17] and [1.18] of Chapter 1. Let us start with equation [1.18] in the practical case where $\sigma_m = 0$:

$$\vec{\nabla} \wedge \underline{\vec{E}} = -\mu \frac{\partial \underline{\vec{H}}}{\partial t}$$

of which the projections onto the Cartesian coordinates are

$$\mu \frac{\partial H_x}{\partial t} = \frac{\partial E_y}{\partial z} - \frac{\partial E_z}{\partial y}$$

$$\mu \frac{\partial H_y}{\partial t} = -\frac{\partial E_x}{\partial z} + \frac{\partial E_z}{\partial x}$$

$$\mu \frac{\partial H_z}{\partial t} = -\frac{\partial E_y}{\partial x} + \frac{\partial E_x}{\partial y}$$

The transformation of these equations is based on the following spectral expression:

$$E_n(x, y, z, t) = \sum_j e_j(z, t) E_n^j(x, y)$$

where $e_i(z, t)$ is an unknown functional coefficient to be defined.

Let the other spectral expansions to be used be

$$E_x = \sum_i e_{x,i}(z,t) E_n^i(x, y) \quad \text{and} \quad E_y = \sum_i e_{y,i}(z,t) E_n^i(x, y)$$

where $e_{x,i}(z, t)$ and $e_{y,i}(z, t)$ are unknown functional coefficients to be defined.

We write the following spectral expansions:

$$H_x = \sum_i h_{x,i}(z,t) E_n^i(x,y)$$

$$H_y = \sum_i h_{y,i}(z,t) E_n^i(x,y)$$

$$H_z = \sum_i h_{z,i}(z,t) E_n^i(x,y)$$

where $h_{x,i}(z, t)$, $h_{y,i}(z, t)$, and $h_i(z, t)$ are unknown functional coefficients to be defined.

We need to expand the partial derivatives of the eigenfunctions E_n^i versus x and y in the Hilbertian base consisting of the E_n^i themselves:

$$\frac{\partial E_n^i}{\partial x} = \sum_j e_{ij}^x E_n^i \quad \text{with} \quad e_{ij}^x = \left(\frac{\partial E_n^i}{\partial x}, E_n^j \right)$$

$$\frac{\partial E_n^i}{\partial y} = \sum_j e_{ij}^y E_n^i \quad \text{with} \quad e_{ij}^y = \left(\frac{\partial E_n^i}{\partial y}, E_n^j \right)$$

These can be written in matrix form:

$$\frac{\partial [E_n]}{\partial x} = [e^x][E_n]$$

$$\frac{\partial [E_n]}{\partial y} = [e^y][E_n]$$

where the infinite dimension square matrices $[e^x]$ and $[e^y]$ are real and antisymmetric.

Integrating by parts gives

$$\iint_{R^2} \frac{\partial E_n^i}{\partial x} E_n^k \, dxdy = -\iint_{R^2} E_n^i \frac{\partial E_n^k}{\partial x} \, dxdy$$

That is to say

$$\left(\frac{\partial E_n^i}{\partial x}, E_n^k\right) = -\left(E_n^i, \frac{\partial E_n^k}{\partial x}\right) = -\left(\frac{\partial E_n^k}{\partial x}, E_n^i\right)$$

and thus $e_{ik}^x = -e_{ki}^x$.

These antisymmetric matrices are invertible because their inverses correspond to the integration operator.

Note that the commutability of the operators $\partial/\partial x$ and $\partial/\partial y$ leads to the matrix commutability

$$[e^x][e^y] = [e^y][e^x] \qquad [3.48]$$

while the eigenfunction equation $HE_n^j = \beta_j^2 E_n^j$ written in matrix form

$$\left(-\frac{\partial^2}{\partial x^2} - \frac{\partial^2}{\partial y^2} - a\right)[E_n] = [\beta^2][E_n] \quad \text{with} \quad \begin{cases} \dfrac{\partial^2[E_n]}{\partial x^2} = [e^x]^2[E_n] \\ \dfrac{\partial^2[E_n]}{\partial y^2} = [e^y]^2[E_n] \end{cases}$$

leads directly to the matrix relation:

$$[e^x]^2 + [e^y]^2 + [a] + [\beta]^2 = [0] \qquad [3.49]$$

The scalar Maxwell's equations become

$$\mu\frac{\partial}{\partial t}[h_x] = \frac{\partial}{\partial z}[e_y] - [e^y][e] \qquad [3.50]$$

$$\mu\frac{\partial}{\partial t}[h_y] = -\frac{\partial}{\partial z}[e_x] + [e^x][e] \qquad [3.51]$$

$$\mu\frac{\partial}{\partial t}[h] = -[e^x][e_y] + [e^y][e_x] \qquad [3.52]$$

Now, we deal with the Maxwell equation

$$\vec{\tilde{\nabla}} \wedge \vec{\tilde{H}} = \sigma \vec{\tilde{E}} + \varepsilon \frac{\partial \vec{\tilde{E}}}{\partial t}$$

The projections onto the Cartesian coordinates are

$$\left(\sigma + \varepsilon \frac{\partial}{\partial t}\right) E_x = -\frac{\partial H_y}{\partial z} + \frac{\partial H_z}{\partial y}$$

$$\left(\sigma + \varepsilon \frac{\partial}{\partial t}\right) E_y = \frac{\partial H_x}{\partial z} - \frac{\partial H_z}{\partial x}$$

$$\left(\sigma + \varepsilon \frac{\partial}{\partial t}\right) E_n = -\frac{\partial H_y}{\partial x} + \frac{\partial H_x}{\partial y}$$

It is very efficient here to use the duality in electromagnetism as follows:

$$\left\{\frac{2}{\mu}[\sigma] + \frac{1}{\mu}[\varepsilon]\frac{\partial}{\partial t}\right\}[e_x] = -\frac{\partial}{\partial z}[h_y] + [e^y][h] \qquad [3.53]$$

$$\left\{\frac{2}{\mu}[\sigma] + \frac{1}{\mu}[\varepsilon]\frac{\partial}{\partial t}\right\}[e_y] = \frac{\partial}{\partial z}[h_x] - [e^x][h] \qquad [3.54]$$

$$\left\{\frac{2}{\mu}[\sigma] + \frac{1}{\mu}[\varepsilon]\frac{\partial}{\partial t}\right\}[e] = [e^x][h_y] - [e^y][h_x] \qquad [3.55]$$

3.2.5. Generalized transmission line matrix equation

Now we gather the six equations of heterogenous stratified media:

$$\mu \frac{\partial}{\partial t}[h_x] = \frac{\partial}{\partial z}[e_y] - [e^y][e]$$

$$\mu \frac{\partial}{\partial t}[h_y] = -\frac{\partial}{\partial z}[e_x] + [e^x][e]$$

$$\mu \frac{\partial}{\partial t}[h] = -[e^x][e_y] + [e^y][e_x]$$

$$\left\{\frac{2}{\mu}[\sigma]+\frac{1}{\mu}[\varepsilon]\frac{\partial}{\partial t}\right\}[e_x]=-\frac{\partial}{\partial z}[h_y]+[e^y][h]$$

$$\left\{\frac{2}{\mu}[\sigma]+\frac{1}{\mu}[\varepsilon]\frac{\partial}{\partial t}\right\}[e_y]=\frac{\partial}{\partial z}[h_x]-[e^x][h]$$

$$\left\{\frac{2}{\mu}[\sigma]+\frac{1}{\mu}[\varepsilon]\frac{\partial}{\partial t}\right\}[e]=[e^x][h_y]-[e^y][h_x]$$

into matrix form. The following square matrices we share into four square sub-matrices. The first is

$$[M]=\begin{bmatrix}[0] & -\mu[1]\\ -\dfrac{1}{\mu}[[\varepsilon]] & [0]\end{bmatrix}$$

where [0] is the zero matrix, [1] is the unity matrix and [[ε]] corresponds to

$$\begin{bmatrix}[\varepsilon] & [0] & [0]\\ [0] & [\varepsilon] & [0]\\ [0] & [0] & [\varepsilon]\end{bmatrix}$$

The second is

$$[N]=\begin{bmatrix}[J] & [0]\\ [0] & [J]\end{bmatrix}$$

where [J] corresponds to

$$\begin{bmatrix}[0] & [1] & [0]\\ -[1] & [0] & [0]\\ [0] & [0] & [0]\end{bmatrix}$$

Matrix [N] is not invertible. The third is

$$[R] = \begin{bmatrix} [\xi] & [0] \\ \dfrac{2}{\mu}[[\sigma]] & [\xi] \end{bmatrix}$$

where $[\xi]$ corresponds to

$$\begin{bmatrix} [0] & [1] & -[e^y] \\ [0] & [0] & [e^x] \\ [e^y] & -[e^x] & [0] \end{bmatrix}$$

and $[[\sigma]]$ corresponds to

$$\begin{bmatrix} [\sigma] & [0] & [0] \\ [0] & [\sigma] & [0] \\ [0] & [0] & [\sigma] \end{bmatrix}$$

Matrix $[R]$ is not invertible. Note that these matrices are not commutable.

Let the column matrix $[u]$ be defined by its transposed row matrix $[u]^T = \begin{bmatrix} [e_x][e_y][e][h_x][h_y][h] \end{bmatrix}$.

In what follows we omit the use of the brackets tied to the matrices in order to simplify the writing. Then, the *generalized transmission line equation* involving an infinite space of functions is as follows:

$$M\frac{\partial u}{\partial t} + N\frac{\partial u}{\partial z} + Ru = 0 \qquad [3.56]$$

3.2.6. *Non-existence of the TM and TE modes separately*

Let us consider the case of the TM mode, the case of the TE mode being deduced by means of duality in electromagnetism. As per the definition, the propagation mode of an electromagnetic wave is called "TM" (respectively "TE") if the longitudinal magnetic field (respectively electrical field) is zero.

We have $[h] = [0]$ in TM mode. From equation [3.52], we obtain

$$[e^x][e_y] - [e^y][e_x] = 0$$

Multiplying equations [3.50] and [3.51] by $[e^x]$ and $-[e^y]$ on their left-hand sides, respectively, then adding them, accounting for the fact that the matrices $[e^x]$ and $[e^y]$ commute [3.48], leads to

$$\mu[e^x]\frac{\partial}{\partial t}[h_x] + \mu[e^y]\frac{\partial}{\partial t}[h_y] = \frac{\partial}{\partial z}\{[e^x][e_y] - [e^y][e_x]\} +$$
$$\{-[e^x][e^y] + [e^y][e^x]\}[e] = [0]$$

This corresponds to the following equation when the initial conditions are zero:

$$[e^x][h_x] + [e^y][h_y] = [0]$$

Note the case of non-zero initial conditions comes back to the previous case because of a translation.

Multiplying on their left-hand sides the two equations [3.53] and [3.54] by $-[e^y]$ and $[e^x]$, respectively, with $[h] = [0]$, then adding them, the differential of the previous linear equation with respect to z is written

$$[e^y]\frac{\partial}{\partial z}[h_y] + [e^x]\frac{\partial}{\partial z}[h_x] = [0]$$
$$= -[e^y]\left\{\frac{2}{\mu}[\sigma] + \frac{1}{\mu}[\varepsilon]\frac{\partial}{\partial t}\right\}[e_x] + [e^x]\left\{\frac{2}{\mu}[\sigma] + \frac{1}{\mu}[\varepsilon]\frac{\partial}{\partial t}\right\}[e_y]$$

From $[e^x][e_y] - [e^y][e_x] = 0$, we obtain $[e_y] = [e^x]^{-1}[e^y][e_x]$

After multiplying by $\mu[e^y]^{-1}$ on the left-hand side, we obtain the following:

$$\left\{2[\sigma] + [\varepsilon]\frac{\partial}{\partial t}\right\}[e_x] = \underbrace{[e^y]^{-1}[e^x]}_{P^{-1}}\underbrace{\left\{2[\sigma] + [\varepsilon]\frac{\partial}{\partial t}\right\}}_{A}\underbrace{[e^x]^{-1}[e^y]}_{P}[e_x]$$

where $[A]$ is an operational matrix that is real and symmetric as it is made up of such matrices and $[P]$ is a real and symmetric matrix as the product of two real antisymmetric matrices.

Multiplying the last equation by the matrix $[P]$ on its left-hand side, we obtain

$$\{[P][A] - [A][P]\}[e_x] = [0]$$

There are two cases: either the matrices $[P]$ and $[A]$ do not commute and we necessarily obtain $[e_x] = [0]$; or these matrices commute and it is possible to obtain $[e_x] \neq 0$: the TM mode can exist. From duality, the TE mode can exist in the same way.

What is the physical meaning of the commutability between $[P]$ and $[A]$?

$$[P][E_n] = [e^x]^{-1}[e^y][E_n] = [e^x]^{-1}\frac{\partial [E_n]}{\partial y} = \int \frac{\partial [E_n]}{\partial y}\, dx$$

then

$$[P][A][E_n] = \int \frac{\partial [A][E_n]}{\partial y}\, dx$$

and the relation

$$\{[P][A] - [A][P]\}[e_x] = [0]$$

becomes

$$\int \frac{\partial [A][E_n]}{\partial y}\, dx = [A]\int \frac{\partial [E_n]}{\partial y}\, dx$$

Differentiation of both terms of the above relation with respect to x allows us to write

$$[E_n] = \frac{\partial [\Phi]}{\partial x}$$

After simplification, this becomes

$$\frac{\partial [A]}{\partial y}\frac{\partial [\Phi]}{\partial x} - \frac{\partial [A]}{\partial x}\frac{\partial [\Phi]}{\partial y} = [0]$$

This means mathematically [VAL1] that there is a relationship between $[A]$ and $[\Phi]$ and then $[E_n]$.

However, physically, there is no relationship between $[A]$, made up of the functions ε and σ depending on the material of the propagation medium, and $[E_n]$

coming from the mathematical properties of the Hamiltonian operator with its boundary conditions. So, we are led to

$$\frac{\partial [A]}{\partial x} = [0] \quad \text{and} \quad \frac{\partial [A]}{\partial y} = [0]$$

which corresponds to

$$\frac{\partial [\varepsilon]}{\partial x} = [0] \quad \frac{\partial [\sigma]}{\partial x} = [0] \quad \frac{\partial [\varepsilon]}{\partial y} = [0] \quad \frac{\partial [\sigma]}{\partial y} = [0]$$

Therefore, the medium is uniform: ε and σ are constant over the whole space. The TM mode can exist alone and, from duality, the TE mode can also exist.

The relations of commutation are because of the uniformity of the propagation media. The TM and TE modes can exist separately in homogenous media only.

3.2.7. *Electrical (or magnetic) currents*

The (electrical or magnetic) vector potential has a fixed direction, the field lines being changed every time. This is the reason why the time domain differential equations (equations [3.22] or [3.38]) and space domain equations (equations [3.34] or [3.39]) of the electrical current across any given surface remain valid without any simplification.

3.3. Conclusion

Maxwell's equations have been converted into integro-differential equations of the electrical or magnetic current in respect of an "electromagnetic Frenet trihedron" linked to vector potential lines in any heterogenous stratified media of which bent flexible interconnection substrates are made. These equations are the basis of the advanced transmission lines theory applied to modern interconnection networks.

In the case of heterogenous but straight rigid stratified media, a full wave analysis from Maxwell's equations to a generalized transmission line matrix equation is possible in any non-bounded domain. This is achieved by means of a spectral analysis similar to the classical one encountered in quantum mechanics applied to semiconductor physics (energy levels computation). This full wave analysis is a rigorous way of modeling multiple dielectric media with lossy conductors.

Chapter 4

Transmission Line Equations

The conversion of Maxwell's equations into integro-differential equations is applied to straight homogenous media of electromagnetic wave propagation that are particular cases of *stratified* electromagnetic media. The modal analysis of the longitudinal fields then of currents and lineic charges is developed by means of the theory of distributions, so gathering the cases of discrete and continuous spectra in order to be compatible with the application of finite element methods for computing the eigenvalues and functions. Finding the transmission line equations of any waveguide completes this.

The propagation mode called *transverse electrical magnetic* (TEM) is presented with the detailed conditions of application including previous spectral considerations. The telegrapher's equation is obtained for current, lineic charges and potentials. The electrostatic behavior of the transversal electrical field is rigorously demonstrated so leading to transmission line equations in matrix form and to the equivalent scheme of coupled lines with lossy dielectric and lossless conductors.

We deal with the application of the *full wave* analysis of straight heterogenous stratified media developed beforehand to the difficult case of lossy parallel conductors in lossy dielectric media. This allows the definition of the so-called Foucault's modes of lossy wave propagation for the longitudinal electrical field. By means of the integration of the main current density within an area where it is constant, we obtain a set of *Foucault's modal* telegrapher's equations for both *modal* current and lineic charges. Assuming the quasi-TEM approximation for the propagation mode, these modal telegrapher's equations become matrix quations related to the modal total currents and electrical (scalar) potentials.

This chapter discusses modeling the effect of bent transmission lines on the electrical behavior by means of the conversion of Maxwell's equations into integro-differential equations previously developed.

4.1. Straight homogenous dielectric media with lossless conductors

4.1.1. *Hypothesis*

– We consider homogenous media made of uniform dielectric permittivity ε, magnetic permeability μ ($= 4\pi \times 10^{-7}$ SI, vacuum value), and electrical conductivity σ.

– Only straight conductors are involved here; thus:

- the radii of curvature and of twisting become infinite,

- letting $\vec{n}\vec{\nabla} = -\dfrac{\partial}{\partial z}$, the integration contour does not depend on z, so

$$\oint \frac{\partial}{\partial z} = \frac{\partial}{\partial z} \oint$$

– The electromagnetic fields along straight lossless conductors have both longitudinal electrical fields, then $E_n \neq 0$, and longitudinal magnetic fields, then $H_n \neq 0$. In the following discussion we point out the analysis of the longitudinal electrical field that is directly applicable to the longitudinal magnetic field by means of the previously demonstrated duality in electromagnetism. Such fields can occur because of:

- any known initial distribution of electromagnetic field,

- or known sources of electromagnetic field set at conductor terminals,

- or known outer radiating sources set in free space far away from conductors.

Let us now define the target of the mathematical modeling in an interconnection network (Figure 2.2). The main technological structures encountered by the shaded conductors in Figure 4.1 in industrial applications are shown in Figures 4.2–4.4.

4.1.2. *Electrical current formulae in TM mode of propagation*

Recall that the basic property of any stratified medium is that ε, σ, μ do not depend on the coordinate along the direction $\vec{n}$ of the vector potential.

The direction of the vector potential is fixed, then the "electromagnetic Frenet triehedron" is fixed, and

$$\frac{\partial}{\partial t}\left[(\vec{e}\vec{\nabla})\right] = 0$$

so that

$$\Re \to \infty, \quad \left[\frac{1}{\Re}\right] = 0, \quad \text{and} \quad \frac{\partial}{\partial t}\left[\frac{1}{\Re}\right] = 0$$

which leads us to write the first integro-differential equation [3.22] encountered in Chapter 3 for the current flow in "stratified media", i.e.

$$\frac{\partial J}{\partial t} = -\oint_{(C_H)} \frac{1}{\mu}\left[(\vec{e}\vec{\nabla}) - \frac{1}{\Re}\right][E_n + (\vec{n}\vec{\nabla})V]\,\mathrm{d}l_H + \oint_{(C_H)} \frac{1}{\mu}\frac{\partial}{\partial t}\left[(\vec{e}\vec{\nabla}) - \frac{1}{\Re}\right]A\,\mathrm{d}l_H$$

as follows:

$$\frac{\partial J}{\partial t} = -\oint_{C_H} \frac{1}{\mu}(\vec{e}\vec{\nabla})[E_n + (\vec{n}\vec{\nabla})V]\,\mathrm{d}l_H = -\oint_{C_H} \frac{1}{\mu}(\vec{e}\vec{\nabla})(\vec{n}\vec{\nabla})V\,\mathrm{d}l_H - \oint_{C_H} \frac{1}{\mu}(\vec{e}\vec{\nabla})E_n\,\mathrm{d}l_H$$

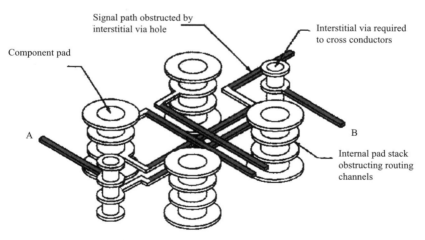

Figure 4.1. *The gray shaded conductors are the targets of the mathematical modeling of the present chapter*

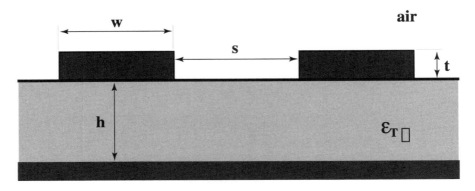

Figure 4.2. *Microstrip structure*

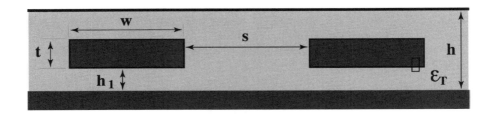

Figure 4.3. *Buried microstrip structure*

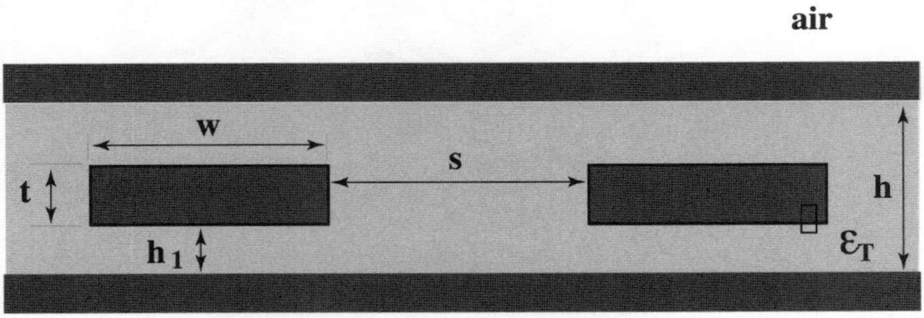

Figure 4.4. *Stripline structure*

Now, $\Re \to \infty$ and the "electromagnetic Frenet trihedron" being fixed, we have the following particular relation:

$$(\vec{e}\vec{\nabla})(\vec{n}\vec{\nabla})V = (\vec{n}\vec{\nabla})(\vec{e}\vec{\nabla})V = -\frac{\partial E_s}{\partial z}$$

where it is recalled that $E_s = -(\vec{e}\vec{\nabla})V$.

Then, we obtain the following equations:

$$\frac{\partial J}{\partial t} = -\frac{1}{\mu}\frac{\partial}{\partial z}\oint_{C_H} E_s \, dl_H - \frac{1}{\mu}\oint_{C_H} (\vec{e}\vec{\nabla})E_n \, dl_H \qquad [4.1]$$

$$\frac{\partial J}{\partial z} = -\left(\sigma + \varepsilon\frac{\partial}{\partial t}\right)\oint_{C_H} E_s \, dl_H \qquad [4.2]$$

4.1.3. *Magnetic current formulae in TE mode of propagation*

4.1.3.1. *TE mode*

In the same way, the equations concerning J_m can be written by using directly the duality between electrical and magnetic currents:

$$\frac{\partial J_m}{\partial t} = -\frac{1}{\varepsilon}\frac{\partial}{\partial z}\oint_{C_E} H_s \, dl_E - \frac{1}{\varepsilon}\oint_{C_E} (\vec{e}_m\vec{\nabla})H_n \, dl_E \qquad [4.3]$$

$$\frac{\partial J_m}{\partial z} = -\mu\frac{\partial}{\partial t}\oint_{C_E} H_s \, dl_E$$

4.1.4. *Spectral analysis of electromagnetic fields*

4.1.4.1. *Lossy propagation equation for the two-dimensional Fourier transform in the space domain*

Let us choose the case of a longitudinal electrical field as a source. The tilde symbol used in $\tilde{E}_n(\beta_x, \beta_y, z, t)$ represents the two-dimensional (2D) Fourier transform of $E_n(x, y, z, t)$ mathematically defined by means of the theory of distributions. We then obtain a spectral representation of the longitudinal electrical field and of the charge density with reverse Fourier transform as:

$$E_n(x,y,z,t) = \frac{1}{4\pi^2} \iint_{R^2} \tilde{E}_n(\beta_x,\beta_y,z,t) e^{j(\beta_x x + \beta_y y)} \, d\beta_x d\beta_y \qquad [4.4]$$

the double integral being extended to the plane orthogonal to field lines of the potential vector going through a point having a coordinate z at time t. A similar spectral representation can be written for the longitudinal magnetic field H_n.

Then, a new lossy propagation equation can be found for the 2D Fourier transform of $\vec{E}$ by means of Maxwell's equations:

$$\nabla^2 E_n - \mu\sigma \frac{\partial E_n}{\partial t} - \mu\varepsilon \frac{\partial^2 E_n}{\partial t^2} = 0 \qquad [4.5]$$

Recalling that

$$\nabla^2 = \frac{\partial^2}{\partial x^2} + \frac{\partial^2}{\partial y^2} + \frac{\partial^2}{\partial z^2}$$

and noting $\beta^2 = \beta_x^2 + \beta_y^2$, two double differentiations behind the integral symbols on the right-hand side of equation [4.4] yield the relation

$$\left(\nabla^2 - \mu\sigma \frac{\partial}{\partial t} - \mu\varepsilon \frac{\partial^2}{\partial t^2}\right) \left[\iint_{R^2} \tilde{E}_n(\beta_x,\beta_y,z,t) e^{j(\beta_x x + \beta_y y)} \, d\beta_x d\beta_y\right]$$

$$= \iint_{R^2} \left[\frac{\partial^2 \tilde{E}_n}{\partial z^2} - \beta^2 \tilde{E}_n - \mu\sigma \frac{\partial \tilde{E}_n}{\partial t} - \mu\varepsilon \frac{\partial^2 \tilde{E}_n}{\partial t^2}\right] e^{j(\beta_x x + \beta_y y)} \, d\beta_x d\beta_y \qquad [4.6]$$

because of [4.5]. This last relation is valid in homogenous media only.

Let us consider three cases involving the electromagnetic field domain in the plane (x, y) bounded or not. In the first case, the electromagnetic waves are traveling in free space as in the "microstrip" structures. In that case, the nullity of the Fourier integral in the second term of [4.6] involves the nullity of the term between brackets, so the "generalized" or *extended telegrapher's equation* becomes

$$\frac{\partial^2 \tilde{E}_n}{\partial z^2} - \beta^2 \tilde{E}_n - \mu\sigma \frac{\partial \tilde{E}_n}{\partial t} - \mu\varepsilon \frac{\partial^2 \tilde{E}_n}{\partial t^2} = 0 \qquad [4.7]$$

4.1.4.2. *Discrete propagation modes in a bounded domain and their "cut-off wavelength"*

In the second case, the electromagnetic waves are traveling in a bounded domain outside of which the field is zero (*waveguide structure*). Then, the boundary conditions of Maxwell's equations involve numerous multiple reflections on the edge of the bounded domain that can be represented only by a denumerable set of solutions of propagation equation [4.5] so that equation [4.7] is not true in this case. Indeed, in order to find these solutions rigorously, we have to note that the finite electromagnetic energy mathematically enforces E_n to have an integrable square.

So, mathematically, E_n belongs to a Hilbert vector space L^2 (consisting of functions) in which the scalar product defined by $\iint fg\, dxdy$ in the plane (x, y) takes place in relation with the *norm* $\|f\|_{L^2} = \sqrt{(f,f)}$ [WAR] as in vector geometry. For instance, the second term of [4.4] represents the scalar product of $\tilde{E}_n$ by $e^{j(\beta_x x + \beta_y y)}$. The advantage of any Hilbert space is to admit a "Hilbertian" base of functions, here denoted $E_n^i(x, y)$, orthogonal, because their scalar product is zero two by two, and to obtain an equivalent to the Pythagorean theorem physically interpretable in terms of energy balance. The development of the electrical field in the Hilbertian base $E_n^i(x, y)$ can be written as

$$E_n(x, y, z, t) = \sum_i e_i(z,t) E_n^i(x, y) \qquad [4.8]$$

Let us impose on E_n^i that it has the Hilbertian norm $\|E_n^i\| = 1$; the energy balance can be expressed by means of the Pythagorean theorem as $\|E_n\|_{L^2}^2 = \sum_i e_i^2$.

The functions E_n^i are the eigenfunctions of the elliptic operator

$$-\frac{\partial^2}{\partial x^2} - \frac{\partial^2}{\partial y^2}$$

related to its eigenvalues as always positive (indeed, let $a = 0$ and $F = E_n^i$ in equation [3.42] in order to obtain a proof) and denoted as β_i^2 respectively define the *TM propagation modes* ($H_n^i = 0$) and are solutions of the equations

$$\frac{\partial^2 E_n^i}{\partial x^2} + \frac{\partial^2 E_n^i}{\partial y^2} = -\beta_i^2 E_n^i \qquad [4.9]$$

Note that the feature $\lambda_i = 2\pi/\beta_i$ is called the *cut-off wavelength of the mode* E_n^i (*tied to harmonic waves only* [LEF] [HAR]) (see section 4.2.3 for computing the lowest cut-off wavelength). The eigenvectors are defined by the boundary conditions $E_n^i = 0$ on the domain edge. The zero eigenvalue leads to the classical Laplace equation and, according to the boundary conditions above, the single possible solution allowed is $E_n^i = 0$ everywhere that defines the so-called TEM mode that is usually considered as a degenerate TM mode.

The longitudinal magnetic field H_n in any bounded domain defines the *TE propagation modes* $E_n^i = 0$ with their corresponding cut-off wavelengths. The eigenvectors are defined by the boundary conditions $\partial H_n^i / \partial v = 0$, v being the coordinate normal to the domain edge.

Note that in processing the affine transformation $\{x = x_0 + kx', y = y_0 + ky'\}$ in equation [4.9], the eigenvalue moduli are proportional to $1/k^2$, the inverse of the square of the similarity ratio k between two similar boundary curves in the plane (x, y). In the general case, the computation of eigenvalues corresponding to realistic technological structures requires the use of up-to-date variational [RAV1] and finite element methods [RAV2], [WAI], which are well documented in the literature.

4.1.4.3. *2D Fourier transform of discrete propagation modes*

Although equation [4.9] leads directly to the extended telegrapher's equation [4.13], it is interesting to obtain a unitary theoretical approach to the two cases of bounded and non-bounded domains in order to bring closer the classical concepts of discrete and continuous spectra. In doing so we can justify the pertinent approximation of a continuous spectrum by a discrete one as regards applying the finite difference method for spectral computations.

Applying the Fourier transform to the two parts of equation [4.9], we obtain

$$(\beta^2 - \beta_i^2)\tilde{E}_n^i(\beta_x, \beta_y) = 0 \qquad [4.10]$$

from which we can easily write $\tilde{E}_n^i(\beta_x, \beta_y) = 0$, which is not interesting physically. Therefore, we are led to find solutions such as $\tilde{E}_n^i(\beta_x, \beta_y) \neq 0$ that can be understood only with the help of the theory of distributions as is rigorously demonstrated here.

Indeed, let $\tilde{E}_n^i$ be a distribution, so a linear mapping on $D(\mathbf{R}^2)$ the vector space of the infinitely derivable functions φ defined on $\mathbf{R}^2$, then $(\beta^2 - \beta_i^2)\tilde{E}_n^i$ is another one. The set of distributions is denoted $D'(\mathbf{R}^2)$. Here, a symbol like $<T, \phi>$ has to be understood by means of the theory of distributions; then it is a linear mapping T from the set of functions ϕ onto the set $\mathbf{R}$ of real numbers. From equation [4.10] we obtain

$$\forall \varphi(\beta_x, \beta_y) \in D(\mathbf{R}^2): \langle (\beta^2 - \beta_i^2)\tilde{E}_n^i, \varphi \rangle = 0$$

We now use polar coordinates; we let angle $\theta = \tan^{-1}(\beta_y / \beta_x)$. It is known that

$$\begin{cases} \langle (\beta^2 - \beta_i^2)\tilde{E}_n^i, \varphi \rangle = \langle \tilde{E}_n^i, (\beta^2 - \beta_i^2)\varphi \rangle \\ \forall \psi(\beta^2, \theta) \in D(\mathbf{R}^2): \langle \delta(\beta^2 - \beta_i^2), \psi(\beta^2, \theta) \rangle = \psi(\beta_i^2, \theta) \end{cases}$$

Let us choose $\psi(\beta^2, \theta) = (\beta^2 - \beta_i^2)\varphi(\beta^2, \theta)$; therefore $\psi(\beta_i^2, \theta) = 0$, and we can write in the space $D'(\mathbf{R}^2)$ of distributions:

$$\langle \tilde{E}_n^i(\beta_x, \beta_y), \psi(\beta^2, \theta) \rangle = 0 \langle \tilde{E}_n^i, (\beta^2 - \beta_i^2)\varphi \rangle$$
$$\forall \Theta(\beta^2, \theta) \in D'(\mathbf{R}^2): \Theta(\beta_i^2, \theta)\psi(\beta_i^2, \theta) = \langle \delta(\beta^2 - \beta_i^2), \Theta(\beta^2, \theta)\psi(\beta^2, \theta) \rangle$$
$$= \langle \Theta(\beta^2, \theta)\delta(\beta^2 - \beta_i^2), \psi(\beta^2, \theta) \rangle = 0$$

Then, there is an infinite number of non-zero distributions

$$\tilde{E}_n^i(\beta_x, \beta_y) = \Theta(\beta^2, \theta)\delta(\beta^2 - \beta_i^2) \qquad [4.11]$$

able to satisfy relation [4.10]. Among them, we have to find the one corresponding to the boundary or initial conditions.

We have to achieve the solution with the eigenfunctions $e_i(z,t)\tilde{E}_n^i(\beta_x, \beta_y)$, where i is directly related to the eigenvalue β_i, and the last result [4.11] can be applied by means of an interpolating distribution $e(\beta^2, z, t)$ so that $e(\beta_i^2, z, t) = e_i(z, t)$. Therefore, solution [4.11] is completed as follows:

$$e_i(z,t)\tilde{E}_n^i(\beta_x, \beta_y) = e(\beta^2, z, t)\Theta(\beta^2, \theta)\delta(\beta^2 - \beta_i^2)$$

Therefore, the Fourier transform of the two parts of [4.8] is expressed as follows:

$$\tilde{E}_n^i(\beta_x,\beta_y,z,t) = \sum_i e_i(z,t)\tilde{E}_n^i(\beta_x,\beta_y) = e(\beta^2,z,t)\Theta(\beta^2,\theta)\sum_i \delta(\beta^2-\beta_i^2) \quad [4.12]$$

This contains the distribution $\sum_i \delta(\beta^2-\beta_i^2)$ which is called the *Dirac comb* distribution and defines the *discrete spectrum* of the 2D Laplace operator in a bounded domain. $\sum_i \delta(\beta^2-\beta_i^2)d(\beta^2)$ defines a spectral measure, $M(\lambda) = \sum_i \gamma(\beta^2-\beta_i^2)$, where γ is the Heaviside function.

4.1.4.4. *Lossy propagation equation in the case of a discrete spectrum in the space domain*

With relation [4.6] and using polar coordinates $\beta_x = \beta\cos\theta$, $\beta_y = \beta\sin\theta$, we obtain

$$d\beta_x d\beta_y = \beta d\beta d\theta = \frac{1}{2}d\beta^2 d\theta$$

which becomes

$$\sum_i \int_0^{+\infty} \left(\frac{\partial^2}{\partial z^2} - \beta^2 - \mu\sigma\frac{\partial}{\partial t} - \mu\varepsilon\frac{\partial^2}{\partial t^2}\right) e(\beta^2,z,t)\delta(\beta^2-\beta_i^2)\Psi(\beta^2,x,y)\,d\beta^2 = 0$$

where

$$\Psi(\beta^2,x,y) = \int_0^{2\pi} \Theta(\beta^2,\theta)e^{j\beta(x\cos\theta+y\sin\theta)}\,d\theta$$

Using the Dirac distribution [1.3] as written by physicists, we obtain

$$\sum_i \Psi(\beta_i^2,x,y)\left(\frac{\partial^2 e_i}{\partial z^2} - \beta_i^2 e_i - \mu\sigma\frac{\partial e_i}{\partial t} - \mu\varepsilon\frac{\partial^2 e_i}{\partial t^2}\right) = 0$$

which has to be valid whatever the functions $\Theta(\beta_i^2,\theta)$ and therefore $\Psi(\beta_i^2,x,y)$. This is possible only if

$$\frac{\partial^2 e_i}{\partial z^2} - \beta_i^2 e_i - \mu\sigma\frac{\partial e_i}{\partial t} - \mu\varepsilon\frac{\partial^2 e_i}{\partial t^2} = 0 \quad [4.13]$$

We deal with the third case corresponding to any domain bounded along one axis only: then the previous developments related to the first two cases have to be applied together. This is the case of a domain between two planes parallel to the plane yz and separated by a distance a for instance ("stripline" structure). The eigenvalues are classically known to be

$$\beta_i^2 = \left(i\frac{\pi}{a}\right)^2$$

the eigenfunctions being either $\sin(\beta_i x)$ for the TM modes or $\cos(\beta_i x)$ for the TE modes, among which there is one called the "basic mode" having the lowest non-zero eigenvalue.

4.1.4.5. *From discrete spectrum to continuous spectrum in the space domain*

In order to propose the use of standard finite element methods in the case of non-bounded structures over free space, we point out the transition from a given bounded domain that is indefinitely growing by similarity from any initial shape to any domain covering the whole space.

Indeed, it has been noted that the eigenvalues β_i^2 become β_i^2/k^2 according to the similarity having the ratio k.

Functional analysis teaches that, i growing indefinitely, the asymptotic behavior of the eigenvalue β_i^2 is like $h^2 i$, where h is a number not depending on i [BRE2].

Let us define the interval I so that

$$i \in I, \quad u_m^2 < \frac{\beta_i^2}{k^2} < u_M^2$$

where u_m and u_M do not depend on k.

As soon as k increases, the number of eigenvalues β_i^2/k^2 within the interval I grows indefinitely which highlights a *continuous spectrum* tied to a non-bounded domain. So, any non-bounded domain can be processed by means of the approximation of its continuous spectrum to a discrete one aimed at numerical methods based on finite elements. Nevertheless, the field has to vanish at infinite distances which is the case according to the law $1/\sqrt{r}$, r being the cylindrical radial coordinate arising from the asymptotic behavior of the Bessel functions discussed in Appendix E.

4.1.5. *Modal analysis of electrical current and lineic charge*

Let us return to the equations

$$\frac{\partial J}{\partial t} = -\frac{1}{\mu}\oint_{C_H} E_s \, dl_H - \frac{1}{\mu}\oint_{C_H}(\vec{e}\vec{\nabla})E_n \, dl_H$$

$$\frac{\partial J}{\partial z} = -\left(\sigma + \varepsilon\frac{\partial}{\partial t}\right)\oint_{C_H} E_s \, dl_H$$

where E_n is non-zero.

The expression of $\oint E_s \, dl_H$ depending on z and t is not equal to zero. By analogy with the *Gauss theorem in electrostatics*, we can write

$$\oint E_s \, dl_H = \frac{Q}{\varepsilon} \qquad [4.14]$$

The value Q introduced here has the same units as an electrical charge per unit length in the direction z inside a perimeter (C_H), depending on z and t. Then, the equations become

$$\frac{\partial J}{\partial t} = -\frac{1}{\mu\varepsilon}\frac{\partial Q}{\partial z} - \frac{1}{\mu}\oint_{C_H}(\vec{e}\vec{\nabla})E_n \, dl_H$$

$$\frac{\partial J}{\partial t} = -\frac{\partial Q}{\partial t} - \frac{\sigma}{\varepsilon}Q \qquad [4.15]$$

We would like to use the spectral expansion according to the Fourier transform

$$E_n(x,y,z,t) = \frac{1}{4\pi^2}\iint_{\mathcal{R}^2}\widetilde{E}_n(\beta_x,\beta_y,z,t)e^{j(\beta_x x+\beta_y y)}d\beta_x d\beta_y$$

Prior to this, it is necessary to transform the second part of $\oint (\vec{e}\vec{\nabla})E_n \, dl_H$ by means of the Green–Riemann theorem transforming any curvilinear integral into a double integral.

According to [3.6], $\vec{e} = \vec{n}\wedge\vec{h}$, we have

$$-(\vec{e}\vec{\nabla})E_n dl_H = (\vec{h}\wedge\vec{n})(\vec{\nabla}E_n)dl_H = (\vec{h}dl_H,\vec{n},\vec{\nabla}E_n)$$

using the classical notations of the mixed product expressed in a Cartesian trihedron (x, y, z) such as

$\vec{h}dl_H$ and its coordinates $(dx, dy, 0)$

$\vec{n}$ and its coordinates $(0, 0, 1)$

$\vec{\nabla}E_n$ and its coordinates $\left(\dfrac{\partial E_n}{\partial x}, \dfrac{\partial E_n}{\partial y}, 0\right)$

Then, we have

$$-(\vec{e}\vec{\nabla})E_n dl_H = \begin{vmatrix} dx & dy & 0 \\ 0 & 0 & 1 \\ \dfrac{\partial E_n}{\partial x} & \dfrac{\partial E_n}{\partial y} & 0 \end{vmatrix} = -\dfrac{\partial E_n}{\partial y}dx + \dfrac{\partial E_n}{\partial x}dy$$

The application of the Green–Riemann formula

$$\oint P\,dx + Q\,dy = \iint \left(\dfrac{\partial Q}{\partial x} - \dfrac{\partial P}{\partial y}\right) dxdy$$

leads to

$$-\oint_{C_H} (\vec{e}\vec{\nabla})E_n dl_H = \iint_{S_H} \left(\dfrac{\partial^2 E_n}{\partial x^2} + \dfrac{\partial^2 E_n}{\partial y^2}\right) dxdy$$

From formula [4.4], by means of the theory of distributions, we obtain the following spectral representation:

$$-\oint_{C_H} (\vec{e}\vec{\nabla})E_n dl_H = -\iint_{R^2} \beta^2 \tilde{E}_n(\beta_x, \beta_y, z, t)\Phi(\beta_x, \beta_y)\,d\beta_x d\beta_y \quad [4.16]$$

where

$$\Phi(\beta_x, \beta_y) = \dfrac{1}{4\pi^2} \iint_{S_H} e^{j(\beta_x x + \beta_y y)}\,dxdy$$

Then, we introduce the feature

$$Q_n = Q_n(z,t) = \varepsilon \oint (\vec{e}\vec{\nabla})E_n(x,y,z,t)\, dl_H \qquad [4.17]$$

which is physically equivalent to a lineic electrical charge. This is a function of z and t, solution of the extended telegrapher's equation. What about the voltage and current?

Let the *telegrapher operator* T be defined by

$$T = \frac{\partial^2}{\partial z^2} - \mu\sigma\frac{\partial}{\partial t} - \mu\varepsilon\frac{\partial^2}{\partial t^2} \qquad [4.18]$$

The *matrix differential operator*

$$[L] = \begin{bmatrix} \mu\varepsilon\dfrac{\partial}{\partial t} & \dfrac{\partial}{\partial z} \\ -\mu\varepsilon\dfrac{\partial}{\partial z} & -\mu\sigma - \mu\varepsilon\dfrac{\partial}{\partial t} \end{bmatrix} \qquad [4.19]$$

gives the equations

$$\begin{cases} \dfrac{\partial J}{\partial t} = -\dfrac{1}{\mu\varepsilon}\dfrac{\partial Q}{\partial z} - \dfrac{1}{\mu}\oint_{C_H}(\vec{e}\vec{\nabla})E_n\, dl_H \\ \dfrac{\partial J}{\partial z} = -\dfrac{\partial Q}{\partial t} - \dfrac{\sigma}{\varepsilon}Q \end{cases} \qquad [4.20]$$

in the following condensed expresion:

$$[L]\begin{bmatrix} J \\ Q \end{bmatrix} = \begin{bmatrix} -Q_n \\ 0 \end{bmatrix} \qquad [4.21]$$

It is easy to check the following relation for the operator $[L]$, a relation coming from the Cayley-Hamilton theorem concerning matrices according to which any matrix $[A]$ having its characteristic polynomial $P(\lambda) = 0$ satisfies the relation $p([A]) = [0]$:

$$[L]^2 + \mu\sigma[L] + \mu\varepsilon T[1] = [0] \qquad [4.22]$$

Multiplying both parts of equation [4.21] on their left-hand side by the matrix $-[L] - \mu\sigma[1]$ gives

$$\mu\varepsilon T[1]\begin{bmatrix} J \\ Q \end{bmatrix} = \{[L] + \mu\sigma[1]\}\begin{bmatrix} Q_n \\ 0 \end{bmatrix} \quad [4.23]$$

that is to say

$$\begin{cases} TJ = \dfrac{\sigma}{\varepsilon}Q_n + \dfrac{\partial Q_n}{\partial t} \\ TQ = -\dfrac{\partial Q_n}{\partial z} \end{cases} \quad [4.24]$$

These are two extended telegrapher's equations with second parts which have to be completed by the following equation without a second part concerning Q_n coming directly from [4.7] and [4.18]:

$$(T - \beta^2)Q_n = 0 \quad [4.25]$$

In the case of a bounded domain, there are different modes E_n^i according to [4.8]. Yet, the theory of distributions through relations [4.12] and [4.16] leads us to write formula [4.17] as follows:

$$Q_n = \sum_i e_i(z,t)Q_n^i \quad \text{with} \quad Q_n^i = \varepsilon\oint_{C_H} (\vec{e}\vec{\nabla})E_n^i \, dl_H \quad [4.26]$$

Indeed, according to

$$-\oint_{C_H} (\vec{e}\vec{\nabla})E_n dl_H = -\iint_{R^2} \beta^2 \tilde{E}_n(\beta_x,\beta_y,z,t)\Phi(\beta_x,\beta_y)\, d\beta_x d\beta_y$$

we have

$$Q_n = \varepsilon \iint_{R^2} \beta^2 \tilde{E}_n(\beta_x,\beta_y,z,t)\Phi(\beta_x,\beta_y)\, d\beta_x d\beta_y$$

and according to

$$\tilde{E}_n^i(\beta_x,\beta_y,z,t) = \sum_i e_i(z,t)\tilde{E}_n^i(\beta_x,\beta_y) = e(\beta^2,z,t)\Theta(\beta^2,\theta)\sum_i \delta(\beta^2 - \beta_i^2)$$

we have

$$Q_n = \varepsilon \iint_{R^2} \beta^2 e(\beta^2, z, t) \Theta(\beta^2, \theta) \sum_i \delta(\beta^2 - \beta_i^2) \Phi(\beta_x, \beta_y) \, \mathrm{d}\beta_x \mathrm{d}\beta_y$$

Using the previously involved polar coordinates leads to

$$Q_n = \frac{\varepsilon}{2} \int_0^\infty \beta^2 e(\beta^2, z, t) \chi(\beta^2) \sum_i \delta(\beta^2 - \beta_i^2) \, \mathrm{d}\beta^2$$

where

$$\chi(\beta^2) = \int_0^{2\pi} \Theta(\beta^2, \theta) \Phi(\beta \cos\theta, \beta \sin\theta) \, \mathrm{d}\theta$$

Using the Dirac distribution (section 1.1.5) as written by physicists, we obtain the first equation [4.26] with

$$Q_n^i = \frac{1}{2} \varepsilon \beta_i^2 \chi(\beta_i^2) \qquad [4.27]$$

Expanding the functions $j_i(z, t)$ and $q_i(z, t)$ as follows:

$$J(z,t) = \sum_i j_i(z,t) Q_n^i \quad \text{and} \quad Q(z,t) = \sum_i q_i(z,t) Q_n^i \qquad [4.28]$$

then equations [4.21] and [4.24] become, respectively

$$\sum_i \left\{ [L] \begin{bmatrix} j_i \\ q_i \end{bmatrix} + \begin{bmatrix} e_i \\ 0 \end{bmatrix} \right\} Q_n^i = \begin{bmatrix} 0 \\ 0 \end{bmatrix}$$

$$\begin{cases} \sum_i \left[T j_i - \left(\frac{\sigma}{\varepsilon} + \frac{\partial}{\partial t} \right) e_i \right] Q_n^i = 0 \\ \sum_i \left(T q_i + \frac{\partial e_i}{\partial z} \right) Q_n^i = 0 \end{cases}$$

which have to be true whatever Q_n^i are related to the functions $E_n^i(x, y)$. This is possible only if

$$[L] = \begin{bmatrix} j_i \\ q_i \end{bmatrix} = \begin{bmatrix} -e_i \\ 0 \end{bmatrix} \qquad [4.29]$$

that is to say

$$\begin{cases} Tj_i = \dfrac{\sigma}{\varepsilon} e_i + \dfrac{\partial e_i}{\partial t} \\ Tq_i = \dfrac{-\partial e_i}{\partial z} \end{cases} \qquad [4.30]$$

We have to complete the last equations with

$$\dfrac{\partial^2 e_i}{\partial z^2} - \beta_i^2 - \mu\sigma \dfrac{\partial e_i}{\partial t} - \mu\varepsilon \dfrac{\partial^2 e_i}{\partial t^2} = 0$$

written as follows:

$$(T - \beta_i^2)e_i = 0 \qquad [4.31]$$

4.1.6. *Modal analysis of scalar and vector potentials*

Let us recall the electrical field E_n relation

$$E_n = -\dfrac{\partial A}{\partial t} + \dfrac{\partial V}{\partial z} \qquad [4.32]$$

arising from [3.9] as regards the hypothesis of [4.11].

Furthermore, the Lorentz condition (section 2.2.2) is written as follows:

$$\vec{\nabla}\vec{A} + \mu\sigma V + \mu\varepsilon \dfrac{\partial V}{\partial t} = 0$$

In a TM mode where the potential vector is in a fixed direction, we have

$$\vec{\nabla}\vec{A} = \vec{\nabla}(\vec{A}\vec{n}) = (\vec{n}\vec{\nabla})A + A(\vec{\nabla}\vec{n})$$

The lines of the potential vector field are straight, so that $\vec{\nabla}\vec{n} = 0$; we obtain

$$-\frac{\partial A}{\partial z} + \mu\sigma V + \mu\varepsilon\frac{\partial V}{\partial t} = 0$$

These relations can be gathered as follows:

$$[L]\begin{bmatrix} A \\ \mu\varepsilon \\ -V \end{bmatrix} = \begin{bmatrix} -E_n \\ 0 \end{bmatrix} \qquad [4.33]$$

The application of the Cayley-Hamilton relation for [L], as discussed above, leads directly to

$$\mu\varepsilon T[1]\begin{bmatrix} A \\ \mu\varepsilon \\ -V \end{bmatrix} = \{[L] + \mu\sigma[1]\}\begin{bmatrix} E_n \\ 0 \end{bmatrix}$$

giving new extended telegrapher's equations with second terms as follows:

$$TV = -\frac{\partial E_n}{\partial z} \qquad [4.34]$$

$$TA = \mu\varepsilon\frac{\partial E_n}{\partial t} + \mu\sigma E_n \qquad [4.35]$$

In the case of a discrete spectrum, we have the following expansions in respect to the formula $E_n(x,y,z,t) = \sum_i e_i(z,t)E_n^i(x,y)$:

$$V(x,y,z,t) = \sum_i v_i(z,t)E_n^i(x,y)$$

$$A(x,y,z,t) = \sum_i a_i(z,t)E_n^i(x,y) \qquad [4.36]$$

Similarly to the previous discussion, we obtain directly

$$[L]\begin{bmatrix} a_i \\ \mu\varepsilon \\ -v_i \end{bmatrix} = \begin{bmatrix} -e_i \\ 0 \end{bmatrix} \qquad [4.37]$$

$$\begin{cases} Tv_i = -\dfrac{\partial e_i}{\partial z} \\ Ta_i = \mu\varepsilon \dfrac{\partial e_i}{\partial t} + \mu\sigma e_i \end{cases} \qquad [4.38]$$

which we have to complete with equations [4.30] and [4.31].

4.1.7. *Transmission line with distributed sources corresponding to a waveguide*

A combination of equations [4.21] and [4.33] so that the unknown longitudinal field E_n is eliminated leads to the transmission line equation without a second term:

$$[L] \begin{bmatrix} J - \dfrac{1}{\mu}\oint_{C_H}(\vec{e}\vec{\nabla})A\,dl_H \\ Q + \varepsilon\oint_{C_H}(\vec{e}\vec{\nabla})V\,dl_H \end{bmatrix} = \begin{bmatrix} 0 \\ 0 \end{bmatrix} \qquad [4.39]$$

Similarly, the three equations

$$\begin{cases} TJ = \dfrac{\sigma}{\varepsilon}Q_n + \dfrac{\partial Q_n}{\partial t} \\ TQ = -\dfrac{\partial Q_n}{\partial z} \end{cases}$$

$$-\dfrac{\partial A}{\partial z} + \mu\sigma V + \mu\varepsilon\dfrac{\partial V}{\partial t} = 0$$

and

$$TV = -\dfrac{\partial E_n}{\partial z}$$

lead to the extended telegrapher's equations:

$$T\left[J - \dfrac{1}{\mu}\oint_{C_H}(\vec{e}\vec{\nabla})A\,dl_H\right] = 0 \qquad [4.40]$$

$$T\left[Q + \varepsilon \oint_{C_H} (\vec{e}\vec{\nabla})V \, dl_H\right] = 0 \qquad [4.41]$$

In the case of a discrete spectrum, we obtain directly the above equations for the modified current and charge:

$$\begin{cases} j_i - \dfrac{1}{\mu} \oint_{C_H} (\vec{e}\vec{\nabla}) a_i \, dl_H \\ q_i + \varepsilon \oint_{C_H} (\vec{e}\vec{\nabla}) v_i \, dl_H \end{cases}$$

These last equations show the way of modeling the effect of any *external electromagnetic perturbation* on the transmission lines provided the scalar and vector potentials or the induced electrical field at the least are known. Anyway, we get here the proof that any waveguide working in a TM mode can be depicted by means of an equivalent transmission line working in the TEM mode but fed with distributed sources defined by scalar and vector potentials.

4.2. TEM mode of wave propagation

4.2.1. *Defining the TEM mode and the transmission lines*

– *TEM mode* is a "degenerate" TM mode defined by a zero longitudinal electrical field, so $E_n = 0$.

– A transmission media center of the propagation of electronic waves in TEM mode is called a "transmission line". This is rigorously achieved when the signal spectrum range corresponds to wavelengths smaller than the "basic cut-off wavelength" of the TE modes of propagation. In contrast to this, when there are outer electromagnetic perturbations for instance, the "equivalent transmission line" corresponding to the waveguide, as highlighted in the previous discussion, can be involved.

– We consider homogenous media for which dielectric permittivity ε, magnetic permeability μ ($4\pi \times 10^{-7}$ SI), and electrical conductibility σ are uniform.

– Conductors are considered straight, so that:
  - the curvature radius $\Re \to \infty$;
  - we write $\vec{n}\vec{\nabla} = -\partial/\partial z$;
  - the integration perimeter does not depend on z, so that $\oint \partial/\partial z = \partial/\partial z \oint$.

4.2.2. Basic existence condition of a TEM propagation mode

$$\frac{\partial^2 E_n^i}{\partial x^2} + \frac{\partial^2 E_n^i}{\partial y^2} = -\beta_i^2 E_n^i$$

Note that the feature $\lambda_i = 2\pi/\beta_i$ is called the *cut-off wavelength of the mode* E_n^i. In what follows, we are interested in the field corresponding to the eigenvalue zero: $i = 0$.

Let us apply the two following vector analysis formulae:

$$\vec{\nabla}(\vec{a} \wedge \vec{b}) = \vec{b}(\vec{\nabla} \wedge \vec{a}) - \vec{a}(\vec{\nabla} \wedge \vec{b})$$

$$\vec{\nabla} \wedge (\vec{\nabla} \wedge \vec{a}) = \vec{\nabla}(\vec{\nabla}\vec{a}) - \nabla^2 \vec{a}$$

which lead to the new formula:

$$(\vec{\nabla} \wedge \vec{a})^2 = \vec{\nabla}[\vec{a} \wedge (\vec{\nabla} \wedge \vec{a})] + (\vec{a}\vec{\nabla})(\vec{\nabla}\vec{a}) - \vec{a}(\nabla^2 \vec{a})$$

which is applied to the vector $\vec{a} = E_n^i \vec{n}$.

As $\partial E_n^0 / \partial z = 0$ and as $\vec{n}$ is a fixed vector, we obtain

$$\nabla^2 \vec{a} = \vec{n}(\nabla^2 E_n^0) = \vec{n}\left(\frac{\partial^2 E_n^0}{\partial x^2} + \frac{\partial^2 E_n^0}{\partial y^2}\right) = \vec{0}$$

Furthermore, $\vec{\nabla}\vec{a} = \vec{\nabla}(E_n^0 \vec{n}) = \partial E_n^0 / \partial z = \vec{0}$ and we write

$$[\vec{\nabla} \wedge (E_n^0 \vec{n})]^2 = \vec{\nabla}\{(E_n^0 \vec{n}) \wedge [\vec{\nabla} \wedge (E_n^0 \vec{n})]\}$$

This is integrated inside the cylindrical domain Ω bounded by the lossless conductive frontier $\partial\Omega$ of the transmission line by means of the divergence theorem transforming a triple integral into a surface one:

$$\iiint_\Omega [\vec{\nabla} \wedge (E_n^0 \vec{n})]^2 \, dv = \iint_{\partial\Omega} \{(E_n^0 \vec{n}) \wedge [\vec{\nabla} \wedge (E_n^0 \vec{n})]\}\vec{e} \, ds$$

Behind the surface integral there is a mixed product denoted as $(\vec{a}, \vec{b}, \vec{c})$, which can be modified in accordance with its classical properties. Indeed

$$\{(E_n^0 \vec{n}) \wedge [\vec{\nabla} \wedge (E_n^0 \vec{n})]\}\vec{e} = (\vec{e}, E_n^0 \vec{n}, \vec{\nabla} \wedge (E_n^0 \vec{n})) = [(\vec{e} \wedge (E_n^0 \vec{n})][\vec{\nabla} \wedge (E_n^0 \vec{n})]$$

However, the vector $\vec{n}$ being constant, the formula $\vec{\nabla} \wedge (\lambda \vec{a}) = (\vec{\nabla}\lambda) \wedge \vec{a} + \lambda(\vec{\nabla} \wedge \vec{a})$ gives

$$\vec{\nabla} \wedge (E_n^0 \vec{n}) = (\vec{\nabla} E_n^0 \vec{n}) \wedge \vec{n}$$

Then, the mixed product is transformed according to $(\vec{a} \wedge \vec{b})(\vec{c} \wedge \vec{d}) = \vec{a}(\vec{b}\vec{c})\vec{d} - \vec{b}(\vec{a}\vec{c})\vec{d}$:

$$[\vec{e} \wedge (E_n^0 \vec{n})][\vec{\nabla} \wedge (E_n^0 \vec{n})] = [\vec{e} \wedge (E_n^0 \vec{n})][(\vec{\nabla} E_n^0) \wedge \vec{n}] = \vec{e}[E_n^0 (\vec{n}\vec{\nabla})E_n^0] - (E_n^0 \vec{n})[\vec{e}\vec{\nabla} E_n^0]\vec{n}$$

Accounting for

$$(\vec{n}\vec{\nabla})E_n^0 = \frac{\partial E_n^0}{\partial z} = 0, \quad (\vec{n})^2 = 1$$

we obtain $E_n^0 (\vec{e}\vec{\nabla})E_n^0$ in the last term.

Thus, we obtain

$$\iiint_\Omega \left[(\vec{\nabla} E_n^0) \wedge \vec{n}\right]^2 dv = -\iint_{\partial\Omega}\left\{E_n^0 (\vec{e}\vec{\nabla})E_n^0\right\} ds = -\iiint_\Omega \vec{\nabla}\left[E_n^0 (\vec{\nabla} E_n^0)\right] dv$$

by means of the divergence theorem again.

The vector formula $\vec{\nabla}(\lambda \vec{a}) = (\vec{\nabla}\lambda)\vec{a} + \lambda(\vec{\nabla}\vec{a})$ yields

$$\vec{\nabla}[E_n^0 (\vec{\nabla} E_n^0)] = (\vec{\nabla} E_n^0)^2 + E_n^0 \underbrace{(\nabla^2 E_n^0)}_{\text{which is zero}}$$

Therefore, we obtain

$$\iiint_\Omega \left\{[(\vec{\nabla} E_n^0) \wedge \vec{n}]^2 + (\vec{\nabla} E_n^0)^2\right\} dv = 0$$

which yields $(\vec{\nabla} E_n^0)^2 = 0$. This requires that each first-order partial derivative is zero, and then the field is uniform. In the case of an electrical field that has to be zero on the lossless conductive boundary, it has to be zero everywhere while the magnetic field remains constant.

So, the eigenvalue zero always corresponds to a uniform field free of any propagation phenomena. The conclusion of this is that a longitudinal electromagnetic field can exist if and only if the signal wavelength range is beyond the lowest cut-off wavelength of the TM or TE mode of propagation.

In microwave theory, it has been accurately demonstrated [BAN] that the lowest frequency is given by the TE mode of propagation then by the longitudinal magnetic field.

4.2.3. *Variational numerical computation of the lowest wavelength*

Let us apply the variational method to the longitudinal magnetic field H_n that has the lowest cut-off wavelength as follows:

– Introduce a test function F belonging to the "Sobolev space H^1" (which means the set of all the functions having a square integral and their derivatives as well).

– Then, multiply the terms of the partial differential equation by F.

– Integrate these terms into the whole domain Ω, then continue to the equation

$$\iint_{R^2} \nabla_s^2 H_n^j F \, dxdy = -\beta_j^2 \iint_{R^2} H_n^j F \, dxdy$$

where

$$\nabla_s^2 = \frac{\partial^2}{\partial x^2} + \frac{\partial^2}{\partial y^2}$$

is the Laplacian operator in the plane (x, y).

The following scalar product is used:

$$(f, g) = \iint_\Omega fg \, dxdy$$

A linear mapping F is defined from the couples (E_n, F) onto the set of real numbers:

$$F(E_n, F) = \iint_\Omega \left(\frac{\partial H_n}{\partial x} \frac{\partial F}{\partial x} + \frac{\partial H_n}{\partial y} \frac{\partial F}{\partial y} - aE_n F \right) dxdy$$

Then, the classical Green's formula:

$$\iint_\Omega \left[\frac{\partial}{\partial x}\left(\frac{\partial H_n}{\partial x}\right) + \frac{\partial}{\partial y}\left(\frac{\partial H_n}{\partial y}\right) \right] F \, dxdy = -\iint_\Omega \left(\frac{\partial H_n}{\partial x}\frac{\partial F}{\partial x} + \frac{\partial H_n}{\partial y}\frac{\partial F}{\partial y} \right) dxdy$$

gives the useful relation $-(\nabla_S^2 H_n^j, F) = F(H_n^j, F)$, leading to the eigenfunctions of the variational equation already met [3.42], $F(H_n^j, F) = \beta_j^2 (H_n^j, F)$, for which we search now for a numerical processing.

In order to do this, we consider a base consisting of a finite number of functions $\{\varphi_i\}$ defining a sub-space of a Hilbert space where the magnetic field H_n is expanded:

$$H_n = \sum_i \varphi_i h_i$$

These functions are chosen with respect to the geometry of the domain where the electromagnetic waves occur. Then, F is equal to one of these functions. F being linear, we can write

$$F(H_n, \varphi_j) = \sum_{i \leq I} \underbrace{F(\varphi_i, \varphi_j)}_{r_{ij} = r_{ji}} h_i$$

The scalar product becomes

$$(H_n, \varphi_j) = \sum_{i \leq I} \underbrace{F(\varphi_i, \varphi_j)}_{m_{ij} = m_{ji}} h_i$$

Defining the one-column matrix $[h]$ having elements h_i, and the symmetric square matrices with dimensions $I \times I$, $[R]$ and $[M]$, having elements r_{ij} and m_{ij}, respectively, we are led to solve the following matrix problem: find the lowest λ so that $[R][h] = \lambda[M][h]$, which means λ is an eigenvalue of the matrix $[M]^{-1}[R]$ which can be diagonalized because it is real and symmetric. The following recursive method is applied.

Let $[h_n] = [R]^{-1}[M][h^*_{n-1}]$ with

$$h^*_{n-1} = \frac{h_{n-1}}{\|h_{n-1}\|_\infty}$$

where $\|h_{n-1}\|_\infty$ is the norm of $[h_{n-1}]$, which is defined as the most equal to the largest element of $[h_{n-1}]$.

Starting from any $[h_0]$, after n times the norm $\left\| [R^{-1}][M][h_n^*] \right\|_\infty$ becomes close to $1/\lambda_1$, λ_1 having the lowest modulus. Knowledge of the cut-off wavelength of all the propagation modes TE and TM comes directly; then, accounting for the propagation velocity $1/\sqrt{\mu\varepsilon}$ of electromagnetic waves, the upper frequency of the signal spectrum traveling in the TEM mode is known.

4.2.4. Telegrapher's equation for current and electrical charge per unit length

Following the assumptions, the equations

$$\frac{\partial J}{\partial t} = -\frac{1}{\mu}\frac{\partial}{\partial z}\oint_{C_H} E_s \, dl_H - \frac{1}{\mu}\oint_{C_H} (\vec{e}\vec{\nabla})E_n \, dl_H \qquad [4.42]$$

$$\frac{\partial J}{\partial z} = -\left(\sigma + \varepsilon\frac{\partial}{\partial t}\right)\oint_{C_H} E_s \, dl_H \qquad [4.43]$$

become

$$\frac{\partial J}{\partial t} = -\frac{1}{\mu}\frac{\partial}{\partial z}\oint_{C_H} E_s \, dl_H \qquad [4.44]$$

$$\frac{\partial J}{\partial z} = -\left(\sigma + \varepsilon\frac{\partial}{\partial t}\right)\oint_{C_H} E_s \, dl_H \qquad [4.45]$$

The expression of $\oint E_s \, dl_H$ depending on z and t is not equal to zero, because if it is zero the previous equations give $J = 0$. By analogy with the *Gauss theorem in electrostatics*, we can write

$$\oint E_s \, dl_H = \frac{Q}{\varepsilon} \qquad [4.45]$$

The value Q introduced here has the same units as an electrical charge per unit length in the direction z inside a perimeter (C_H), depending on z and t, so

$$\frac{\partial J}{\partial t} = -\frac{1}{\mu\varepsilon}\frac{\partial Q}{\partial z} \qquad [4.46]$$

In the same way, the formulation of the current derivative relative to space becomes

$$\frac{\partial J}{\partial z} = -\left(\sigma + \varepsilon \frac{\partial}{\partial t}\right) \oint_{C_H} E_s \, dl_H \qquad [4.47]$$

Applying the Gauss theorem, we obtain

$$\frac{\partial J}{\partial z} = -\frac{\partial Q}{\partial t} - \frac{\sigma}{\varepsilon} Q \qquad [4.48]$$

So, we can deduce

$$[4.46] \Rightarrow \frac{\partial^2 J}{\partial z^2} = -\frac{1}{\mu\varepsilon} \frac{\partial^2 Q}{\partial t \partial z}$$

$$[4.48] \Rightarrow \frac{\partial^2 J}{\partial z^2} = -\frac{\partial^2 Q}{\partial z \partial t} - \frac{\sigma}{\varepsilon} \frac{\partial Q}{\partial z} = \mu\varepsilon \frac{\partial^2 J}{\partial t^2} + \mu\sigma \frac{\partial J}{\partial t}$$

We are led to the *telegrapher's equation for electrical current*:

$$\frac{\partial^2 J}{\partial z^2} - \mu\sigma \frac{\partial J}{\partial t} - \mu\varepsilon \frac{\partial^2 J}{\partial t^2} = 0 \qquad [4.49]$$

In the same way, [4.46] and [4.48] lead to

$$\frac{\partial^2 J}{\partial z \partial t} = -\frac{1}{\mu\varepsilon} \frac{\partial^2 Q}{\partial z^2} = \frac{\partial^2 Q}{\partial t^2} + \frac{\sigma}{\varepsilon} \frac{\partial Q}{\partial t}$$

We are led to the *telegrapher's equation for electrical charge per unit length*:

$$\frac{\partial^2 Q}{\partial z^2} - \mu\sigma \frac{\partial Q}{\partial t} - \mu\varepsilon \frac{\partial^2 Q}{\partial t^2} = 0 \qquad [4.50]$$

4.2.5. *Lorentz condition and telegrapher's equation for vector potentials and scalars in TEM mode*

Let us recall the Lorentz condition, written as follows:

$$\vec{\nabla}\vec{A} + \mu\sigma V + \mu\varepsilon\frac{\partial V}{\partial t} = 0 \qquad [4.51]$$

In a TM mode where the potential vector is in a fixed direction, we have

$$\vec{\nabla}\vec{A} = \vec{\nabla}(\vec{A}\vec{n}) = (\vec{n}\vec{\nabla})A + A(\vec{\nabla}\vec{n}) = (\vec{n}\vec{\nabla})A = -\frac{\partial A}{\partial z}$$

Because the lines of the potential vector field are straight, so that $\vec{\nabla}\vec{n} = 0$, we obtain

$$-\frac{\partial A}{\partial z} + \mu\sigma V + \mu\varepsilon\frac{\partial V}{\partial t} = 0 \qquad [4.52]$$

Because of the TEM mode, the longitudinal electrical field (in the direction z) E_n is equal to zero, so the equation

$$E_n = -\frac{\partial A}{\partial t} - (\vec{n}\vec{\nabla})V$$

gives

$$\frac{\partial A}{\partial t} + \frac{\partial V}{\partial z} = 0 \qquad [4.53]$$

Equations [4.52] and [4.53] form a system of transmission line equations for the vector and scalar potentials.

Accounting for [4.52] and [4.53], we obtain

$$\frac{\partial^2 A}{\partial z^2} = -\mu\sigma\frac{\partial V}{\partial z} - \mu\varepsilon\frac{\partial^2 V}{\partial z\partial t} = -\mu\sigma\left(-\frac{\partial A}{\partial t}\right) - \mu\varepsilon\left(-\frac{\partial^2 A}{\partial t^2}\right)$$

from which the *telegrapher's equation for the modulus of the vector potential in a fixed direction* is obtained:

$$\frac{\partial^2 A}{\partial z^2} - \mu\sigma\frac{\partial A}{\partial t} - \mu\varepsilon\frac{\partial^2 A}{\partial t^2} = 0 \qquad [4.54]$$

In the same way, arranging [4.50] with [4.51], we obtain

$$\frac{\partial^2 V}{\partial z^2} = -\frac{\partial^2 A}{\partial z \partial t} = \mu\sigma\frac{\partial V}{\partial t} + \mu\varepsilon\frac{\partial^2 V}{\partial t^2}$$

leading to the *telegrapher's equation for the scalar potential*:

$$\frac{\partial^2 V}{\partial z^2} - \mu\sigma\frac{\partial V}{\partial t} - \mu\varepsilon\frac{\partial^2 V}{\partial t^2} = 0 \qquad [4.55]$$

4.2.6. Lineic distribution of electrical charges and the Poisson equation

The Maxwell-Ampere equation $\vec{\nabla} \wedge \vec{H} = \vec{j}$ leads to the classical equation of continuity:

$$\vec{\nabla}\vec{j} = 0 \qquad [4.56]$$

According to the basic relation

$$\vec{j} = \left(\sigma + \varepsilon\frac{\partial}{\partial t}\right)\vec{E}$$

this equation can be written as

$$\vec{\nabla}\left[\left(\sigma + \varepsilon\frac{\partial}{\partial t}\right)\vec{E}\right] = 0$$

Developing this, we obtain

$$\left(\sigma + \varepsilon\frac{\partial}{\partial t}\right)\vec{\nabla}\vec{E} + \left[(\vec{\nabla}\sigma) + (\vec{\nabla}\varepsilon)\frac{\partial}{\partial t}\right]\vec{E} = 0 \qquad [4.57]$$

which is a general equation for heterogenous media.

In homogenous media, $\vec{\nabla}\sigma = 0$ and $\vec{\nabla}\varepsilon = 0$.

The volumetric density of charges q is connected to the electrical field by the Poisson equation (3D):

$$\vec{\nabla}\vec{E} = \frac{q}{\varepsilon} \qquad [4.58]$$

The general equation of continuity [4.57] can be reduced to

$$\sigma q + \varepsilon \frac{\partial q}{\partial t} = 0$$

the solution of which being $q = q_0 e^{-(\sigma/\varepsilon)t}$ becomes zero all the faster as σ becomes higher or ε lower. In lossless dielectric media $\sigma = 0$ and the charge density remains non-zero and constant.

Thus, when electrical charges are not initially introduced, the charge density inside homogenous media remains zero.

The charge Q per unit length is then concentrated on the closed curve (C) boundary of the homogenous domain with a surface density Q_C such that:

$$Q = \oint_C Q_C \, dl_C \qquad [4.59]$$

where dl_C is an elementary arc of (C).

According to $\oint E_s \, dl_H = Q/\varepsilon$, we note that the electrical field E_s on (C) is equal to Q_C/ε and its direction is orthogonal to the curve (C) on which the surface density Q_C is distributed.

Bearing in mind the theory of distributions (equation [1.5]), we can introduce the Dirac distribution $\delta(C)$:

$$\oint_C Q_C \, dl_C = \iint_S Q_C \delta(C) \, dS \qquad [4.60]$$

where S is the domain inside (C).

With the equation $\oint E_s \, dl_H = Q/\varepsilon$ defining Q, the expression $E_s dl_H$ appears and can be written according to [3.10]:

$$E_s dl_H = -(\vec{e}\vec{\nabla})V dl_H$$

According to [3.8]

$$E_s dl_H = (\vec{h} \wedge \vec{n})(\vec{\nabla} V) dl_H = (\vec{h} dl_H, \vec{n}, \vec{\nabla} V)$$

using the classical notations of the mixed product expressed in a Cartesian trihedron (x, y, z) such as

$\vec{h} dl_H$ and its coordinates $(dx, dy, 0)$

$\vec{n}$ and its coordinates $(0, 0, 1)$

$\vec{\nabla} V$ and its coordinates $\left(\dfrac{\partial V}{\partial x}, \dfrac{\partial V}{\partial y}, 0 \right)$

Then, we have

$$E_s dl_H = \begin{vmatrix} dx & dy & 0 \\ 0 & 0 & 1 \\ \dfrac{\partial V}{\partial x} & \dfrac{\partial V}{\partial y} & 0 \end{vmatrix} = -\dfrac{\partial V}{\partial y} dx + \dfrac{\partial V}{\partial x} dy$$

The application of the Green-Riemann formula

$$\oint P\, dx + Q\, dy = \iint \left(\dfrac{\partial Q}{\partial x} - \dfrac{\partial P}{\partial y} \right) dxdy$$

leads to

$$\oint E_s\, dl_H = \iint_S \left(\dfrac{\partial^2 V}{\partial x^2} + \dfrac{\partial^2 V}{\partial y^2} \right) dxdy \qquad [4.61]$$

The equation $\oint E_s\, dl_H = Q/\varepsilon$ then becomes

$$\iint_S \left(\dfrac{\partial^2 V}{\partial x^2} + \dfrac{\partial^2 V}{\partial y^2} \right) dxdy = \dfrac{Q}{\varepsilon} = \iint_S \dfrac{Q_c}{\varepsilon} \delta(C)\, dxdy$$

according to [4.59] and [4.60].

Denoting ∇_S^2 as the *2D Laplace operator*

$$\frac{\partial^2}{\partial x^2} + \frac{\partial^2}{\partial y^2}$$

we deduce the Poisson equation for the scalar potential:

$$\nabla_s^2 V = \frac{Q_C}{\varepsilon} \delta(C) \qquad [4.62]$$

Analytical formulae suitable for classical technological structures can be found in Appendix C.

4.2.7. *Transmission line equations for lossy dielectrics and lossless conductors*

The previous equation differs from the usual Poisson equation in electrostatics in that V and Q_C depend on z and t. Nevertheless, the electrical field E_s is orthogonal to the curve (C) which is called an *equipotential line*.

In order to keep the electrical field E_s different from zero, it is necessary, according to the Poisson equation above, that any boundary curve like (C) be multiply connected which means consisting of at least two closed plane curves having no common point.

So, the Poisson equation has a single solution corresponding to the classical Dirichlet problem defined by the following boundary condition: two different known potentials on at least two closed curves on the same plane without any common point.

In particular, arbitrarily imposing an electrical potential equal to 1 V on one of the curves and 0 V on the other curves defines the elements C_{ij} of the *capacitance matrix per unit length* $[C_{ij}]$, as the total electrical charge on curve i set at the electrical potential = 1 V when the potential of the other curves $j \neq i$ is zero. See Appendix C for some analytical approaches to the matrix elements related to some basic technological structures.

If the electrical potential of this curve has any value V_i, the following total charge results: $Q_i = C_{ii} V_i$. When the potential V of the other curves are not zero any more, we obtain electrical charges $Q_i = \sum_i C_{ij} V_i$ that can be written in matrix form:

$$[Q] = [C][V] \qquad [4.63]$$

In this section, we write the expressions of the electrical charge with sign "+" instead of sign "-" for the sake of simplicity as

$$\frac{\partial J}{\partial t} = \frac{1}{\mu}\frac{\partial Q}{\partial z} \text{ and } \frac{\partial J}{\partial z} = \frac{\partial Q}{\partial t} + \frac{\sigma}{\varepsilon}Q$$

which becomes

$$\frac{\partial [J]}{\partial z} = [C]\frac{\partial [V]}{\partial t} + \frac{\sigma}{\varepsilon}[C][V] \qquad [4.64]$$

$$\frac{\partial [V]}{\partial z} = \mu\varepsilon[C]^{-1}\frac{\partial [J]}{\partial t} \qquad [4.65]$$

According to the Faraday-Lenz theorem, equation [4.65] contains the expression $\mu\varepsilon[C]^{-1}$ that can be defined as the *inductance matrix per unit length L*:

$$[L] = \mu\varepsilon[C]^{-1} \qquad [4.66]$$

This last formula is only applicable in homogenous media (μ and ε strictly uniform). According to Ohm's law, the equation gives rise to the matrix $(\sigma/\varepsilon)[C]$ which is the *conductance matrix per unit length [G]*:

$$[G] = \frac{\sigma}{\varepsilon}[C] \qquad [4.67]$$

Finally, equations [4.64] and [4.65] can be written as

$$\frac{\partial [J]}{\partial z} = [C]\frac{\partial [V]}{\partial t} + [G][V] \qquad [4.68]$$

$$\frac{\partial [V]}{\partial z} = [L]\frac{\partial [J]}{\partial t} \qquad [4.69]$$

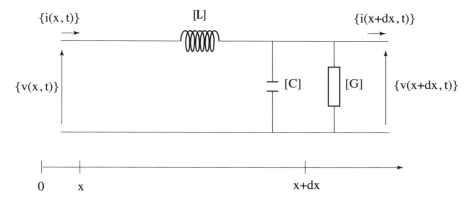

Figure 4.5. *Electrical scheme of a lossy dielectric transmission line*

In accordance with the usual presentation of the transmission lines theory initiated by Lord Kelvin in 1884, the system consists of coupled transmission line equations in the case of lossy dielectrics and lossless conductors.

So, equations [4.68] and [4.69] correspond to the electrical scheme shown in Figure 4.5.

Let us point out the effect of outer electromagnetic perturbations assuming the quasi-TEM mode of propagation according to which we can write $q_i = Cv_i$. In this case, the perturbed transmission line can be depicted by the scheme shown in Figure 4.6.

4.2.8. *Green's kernels and the numerical computation of lineic parameters*

The Poisson equation [4.62] for the scalar potential

$$\nabla_s^2 V = \frac{Q_C}{\varepsilon} \delta(C)$$

is a linear equation that we have to solve by taking advantage of the concept of kernel distribution of a linear mapping reviewed in section 1.1.5.6 devoted to the theory of distributions. Numerous methods, based on using Green's function for solving the Poisson equation, have been widely published in the literature [DUR] [WEI] [SCHE] [VEN] [AHM].

We have already met linear equations that can be written as $A^{-1}(u) = f$ (here $A^{-1}(u) = \nabla_s^2$) and it has been shown that if there is a linear mapping B commuting with A^{-1}, then the solution is $u = B(G)$, G being an elementary solution of $A^{-1}(u) = \delta$ and called the Green's kernel of this equation (here the Poisson equation).

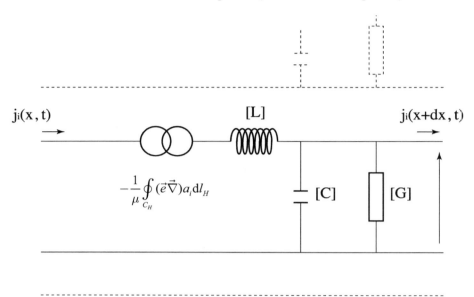

Figure 4.6. *Electrical scheme of a transmission line perturbed by an external electromagnetic field defined by scalar and vector potentials according to the quasi-TEM approximation*

In our application, the *Green's kernel* is physically understood as the electrostatic influence of a straight line on cylindrical conductors set at zero potential. In the present case, A^{-1} commutes with the translations because the linear operator $A^{-1} = \nabla_S^2$ has constant coefficients which means the physical law has a "symmetry translation" (section 1.1.3), then B can be chosen as the linear operator of the convolution product of the function f with the Dirac distribution. Then the solution becomes in this case a convolution product of f with the Green's kernel $u = f * G$.

In electromagnetism, as in any other domain of physics, the features involved have a finite energy so justifying the mathematical models based on the use of Hilbert spaces fed by the 2D scalar product of functions like f and g defined by $\iint_S fg \, dS$, S being the cross-section of the interconnection media, the boundary of which necessarily forms a multiply connected curve. Let a Hilbertian base be such that

$$(\varphi_n, \varphi_m) = 0 \quad \text{if} \quad n \neq m$$
$$(\varphi_n, \varphi_m) = 1 \quad \text{if} \quad n = m$$

Then, we obtain the following Hilbertian expansion where we point out separately the variables (x', y', z'), hereafter denoted x', which are involved in the scalar products, and those (x, y, z), hereafter denoted x, which are not:

$$u(x) = \sum_n (u(x'), \varphi_n(x'))\varphi_n(x) \quad \text{and} \quad f(x) = \sum_n (f(x'), \varphi_n(x'))\varphi_n(x)$$

Closure relation [2.16] was shown in section 2.2.3:

$$\delta(x - x_0) = \sum_n \varphi_n(x)\varphi_n(x_0)$$

This closure relation is useful for finding the Green's kernels corresponding to given constraints and therefore all the solutions of linear problems in so far as the Hilbertian base is made up of the eigenfunctions of $A^{-1} = \nabla_S^2 : \nabla_S^2 \varphi_n = \lambda_n \varphi_n$ if A^{-1} meets the mathematical conditions of compactness [BRE2]. This compactness is naturally met in any interconnection media since it is assumed that the domain of electromagnetic wave propagation is bounded by the frontier between the internal dielectric and the external atmosphere considered as a vacuum. If this is not the case, A^{-1} does not meet the conditions of compactness, which involves the set of eigenfunctions that is not denumerable; in other words, the spectrum of eigenvalues λ is continuous, so made of compact segments of the real straight line.

In the case of compactness, we obtain the *Green's kernel* in an explicit form well suited for meeting the boundary conditions to which the eigenvalues and functions correspond:

$$G(x, x') = G(x - x') = \sum_n \frac{1}{\lambda_n} \varphi_n(x)\varphi_n(x')$$

which is symmetric.

Then, the following Hilbertian expansion:

$$\frac{1}{\varepsilon} Q_C(x)\delta(C) = \sum_n \left(\frac{1}{\varepsilon} Q_C(x')\delta(C), \varphi_n(x') \right) \varphi_n(x)$$

leads directly to the solution:

$$V(x) = \sum_n \frac{1}{\lambda_n} \left(\frac{1}{\varepsilon} Q_C(x')\delta(C), \varphi_n(x') \right) \varphi_n(x)$$

$$= \sum_n \frac{1}{\lambda_n} \iint_S \left(\frac{1}{\varepsilon} Q_C(x')\delta(C), \varphi_n(x') \right) \mathrm{d}S \varphi_n(x)$$

$$= \sum_n \frac{1}{\lambda_n} \oint_C \left[\frac{1}{\varepsilon} Q_C(x'), \varphi_n(x') \right] \mathrm{d}l \varphi_n(x) = \oint_C \underbrace{\left[\sum_n \frac{1}{\lambda_n} \varphi_n(x'), \varphi_n(x) \right]}_{G(x,x')=G(x-x')} \frac{1}{\varepsilon} Q_C(x') \, \mathrm{d}l$$

which yields the convolution product:

$$V(x) = \oint_C G(x-x') \frac{1}{\varepsilon} Q_C(x') \, \mathrm{d}l = \frac{1}{\varepsilon} Q_C * G$$

This is an integral equation to be solved with regard to the unknown charge lineic density when the scalar potential is known along the whole necessarily multiply connected curve.

In the case of non-compactness of $A^{-1} = \nabla_S^2$, which means physically the domain of the electrostatic field cannot be bounded, there is a continuous spectrum of eigenvalues and it has been shown in section 4.1.4.5 the convergence in the sense of distributions of the discrete expansions depending on the integer n towards integral forms involving a *spectral measure* $S(\lambda)$:

$$\delta(x-x') = \int \varphi_\lambda(x) \varphi_\lambda(x') \, \mathrm{d}[S(\lambda)]$$

$$G(x-x') = \int \frac{1}{\lambda} \varphi_\lambda(x) \varphi_\lambda(x') \, \mathrm{d}[S(\lambda)]$$

$$f(x) = \int (f(x'), \varphi_\lambda(x')) \varphi_\lambda(x) \, \mathrm{d}[S(\lambda)]$$

where f is either V or $(1/\varepsilon) Q_C(x)\delta(C)$. These relations lead again to the same integral equation.

In the case of free space, which again involves the criterion of non-compactness, it is assumed that the electrical charge density is gathered on a single axis: then, the elementary known Green's function is equal to $Ln(x)$, x being the distance to the axis. If there is a ground plane parallel to this axis, the Green's function, estimated by the method of *electrostatic images* similar to that presented in section 2.1.2 on the reflection of electromagnetic fields, becomes $Ln(x-x_0) + Ln(x+x_0)$, x_0 being the distance between the plane and the axis.

This work deals with numerical computation through sharing each single connected curve into small arcs where the Green's function and the electrical charge lineic density can be considered constant in accordance with the required accuracy of the computation: so is the *discretization process of the integral equation*.

Let

$$\underbrace{C}_{\substack{\text{multiply}\\\text{connected}\\\text{curves}}} = \bigcup_i \underbrace{C_i}_{\substack{\text{single}\\\text{connected}\\\text{curves}}} = \bigcup_i \left(\bigcup_j \underbrace{C_{i,j}}_{\substack{\text{elementary}\\\text{arcs of curve}}} \right)$$

setting $x = x_{i,j} \in C_{i,j}$, and assuming each elementary arc length is so small that the Green's function and the electrical charge lineic density can be considered constant in accordance with the required accuracy. This yields the following linear equations where the electrical charge lineic density at the points $x = x_{i,j} \in C_{i,j}$ has to be computed knowing the scalar potential at these points:

$$\underbrace{V(x_{k,l})}_{\text{known}} = \sum_{i,j} \underbrace{\frac{1}{\varepsilon} G(x_{k,l} - x'_{i,j}) l_{i,j}}_{\text{known}} \underbrace{Q_C(x'_{i,j})}_{\text{unknown}}$$

which can be solved by any classical numerical method (Appendix D).

Of course, all the arcs belonging to the same single connected conductor have the same scalar potential: $V(x_{k,l})$ do not depend on the integer index l.

Furthermore, the total electrical charge of each single connected conductor is given by

$$Q_i = \sum_j Q_C(x'_{i,j})$$

Thus, knowing all the $Q_C(x'_{i,j})$ then the Q_i, it remains to account for the relations $Q_i = \sum_k C_{i,k} V(x_{k,l})$ which deal with the computation of $C_{i,k}$ by setting

$$V(x_{k,l}) = 0 \quad \text{if} \quad k \neq i \quad \text{and} \quad V(x_{i,l}) = 1$$

Let us recall that the above computations are valid only for interconnection media having homogenous dielectrics.

4.3. Quasi-TEM approximation for lossy conductors and dielectrics

4.3.1. *Foucault's modal currents of electromagnetic field propagation in lossy media*

Let us come back to the situation depicted in section 3.2.2.

Everywhere inside any whole heterogenous medium, the following relation is valid:

$$j = \sigma_c E_n + \varepsilon_c \frac{\partial E_n}{\partial t}$$

from which we obtain the total current flowing across S_H, being the cross-sectional area within the boundary C_H between the lossy conductor and the dielectric as shown in Figure 3.6:

$$J = \iint_{S_H} \left(\sigma_c E_n + \varepsilon_c \frac{\partial E_n}{\partial t} \right) dxdy$$

According to Figure 3.6, it is assumed that the electrical permittivity and conductibility are constant in any inner area closed to S_H; these are denoted as ε_c and σ_c, respectively, hereafter. In this case, the Fourier transform defined by

$$E_n(x,y,z,t) = \frac{1}{4\pi^2} \iint_{R^2} \tilde{E}_n(\beta_x, \beta_y, z, t) e^{j(\beta_x x + \beta_y y)} \, d\beta_x d\beta_y$$

can be involved successfully for going from the electromagnetic field spectral representation to that of the total current inside S_H since ε_c and σ_c are constant. Indeed, after an exchange of integrations, we obtain

$$J(z,t) = \left(\sigma_c + \varepsilon_c \frac{\partial}{\partial t} \right) \iint_{R^2} \tilde{E}_n(\beta_x, \beta_y, z, t) \Phi(\beta_x, \beta_y) \, d\beta_x d\beta_y$$

as was done at the beginning of this chapter. Having a discrete spectrum according to equations [3.41] or [3.42], the equation

$$\tilde{E}_n(\beta_x, \beta_y, z, t) = \sum_i e_i(z,t) \tilde{E}_n^i(\beta_x, \beta_y) = e(\beta^2, z, t) \Theta(\beta^2, \theta) \sum_i \delta(\beta^2 - \beta_i^2)$$

is applicable as follows:

$$J(z,t) = \left(\sigma_c + \varepsilon_c \frac{\partial}{\partial t}\right) \sum_i \iint_{R^2} e(\beta^2,z,t)\Theta(\beta^2,\theta)\delta(\beta^2 - \beta_i^2)\,\mathrm{d}\beta_x \mathrm{d}\beta_y$$

Using polar coordinates and physicists' writing of the Dirac distribution in a way similar to that to obtain the relation

$$Q_n^i = \frac{1}{2}\varepsilon\beta_i^2 \chi(\beta_i^2)$$

we can write

$$J(z,t) = \left(\frac{\sigma_c}{2} + \frac{\varepsilon_c}{2}\frac{\partial}{\partial t}\right) \sum_i \chi(\beta_i^2) e_i(z,t)$$

From

$$Q_n = \sum_i e_i(z,t) Q_n^i \quad \text{with} \quad Q_n^i = \varepsilon \oint_{C_H} (\vec{e}\vec{\nabla}) E_n^i \, \mathrm{d}l_H$$

we obtain

$$\chi(\beta_i^2) = \frac{2Q_n^i}{\varepsilon_c \beta_i^2}$$

and from the formulae $J(z,t) = \sum_i j_i(z,t) Q_n^i$ and $Q(z,t) = \sum_i q_i(z,t) Q_n^i$ of the spectral expansion of the electrical current, the last equation gives

$$J(z,t) = \frac{1}{\beta_i^2}\left(\frac{\sigma_c}{\varepsilon_c} + \frac{\partial}{\partial t}\right) e_i(z,t) \qquad [4.70]$$

This relation defines the Foucault modal currents.

This leads us to express the functions e_i in the equation

$$[L]\begin{bmatrix} j_i \\ q_i \end{bmatrix} = \begin{bmatrix} -e_i \\ 0 \end{bmatrix}$$

Then using the inverse of the linear operator

$$\frac{\sigma_c}{\varepsilon_c} + \frac{\partial}{\partial t}$$

we obtain

$$\begin{bmatrix} \beta_i^2 \left(\frac{\sigma_c}{\varepsilon_c} + \frac{\partial}{\partial t} \right)^{-1} + \mu_0 \varepsilon_d \frac{\partial}{\partial t} & \frac{\partial}{\partial z} \\ -\mu_0 \varepsilon_d \frac{\partial}{\partial z} & -\mu_0 \sigma_d - \mu_0 \varepsilon_d \frac{\partial}{\partial t} \end{bmatrix} \begin{bmatrix} j_i \\ q_i \end{bmatrix} = \begin{bmatrix} 0 \\ 0 \end{bmatrix} \quad [4.71]$$

This consists of many separate transmission line matrix equations, each one related to each Foucault mode of electromagnetic field propagation.

4.3.2. *Quasi-TEM approximation of coupled lossy transmission lines*

Assuming the quasi-TEM approximation, that is, corresponding to a weak electrical inside the slightly lossy conductors involved in coupled transmission lines, the following relations are applied:

$$\mu_0 \sigma_d = [L][G] = [G][L]$$

$$\mu_0 \varepsilon_d = [L][C] = [C][L]$$

Using the matrix blocks process and the above relations, equation [4.71] becomes easily

$$\frac{\partial [v_i]}{\partial z} = -[L]\frac{\partial [j_i]}{\partial t} - [C]^{-1} \beta_i^2 \left(\frac{\sigma_c}{\varepsilon_c} + \frac{\partial}{\partial t} \right)^{-1} [j_i] \quad [4.72]$$

$$\frac{\partial [j_i]}{\partial z} = -[C]\frac{\partial [v_i]}{\partial t} - [G][v_i] \quad [4.73]$$

The second equation, devoted to the dielectric losses, is similar to [4.68]. In contrast, the first equation highlights the following new expression, not encountered in [4.69], that we expand into operational series with respect to $\partial/\partial t$ limited to the two first physically significant terms in the applications:

$$[C]^{-1}\beta_i^2\left(\frac{\sigma_c}{\varepsilon_c}+\frac{\partial}{\partial t}\right)^{-1}[j_i]=\frac{\varepsilon_c\beta_i^2}{\sigma_c}[C]^{-1}[j_i]-\frac{\varepsilon_c^2\beta_i^2}{\sigma_c^2}[C]^{-1}\frac{\partial[j_i]}{\partial t}+\cdots$$

The first term on the right-hand side corresponds to Foucault's modal lineic resistors:

$$[R_i]=\frac{\varepsilon_c\beta_i^2}{\sigma_c}[C]^{-1} \qquad [4.74]$$

while the second term corresponds to Foucault's modal lineic inductances linked to the conductor losses so that the lineic inductance of lossless transmission lines becomes in the case of conductor losses:

$$[L]\rightarrow[L_i]=[L]-\frac{\varepsilon_c^2\beta_i^2}{\sigma_c^2}[C]^{-1}=\left(\mu_0\varepsilon_d-\frac{\varepsilon_c^2\beta_i^2}{\sigma_c^2}\right)[C]^{-1} \qquad [4.75]$$

Finally, equation [4.71] becomes the general coupled lossy transmission line equation:

$$\frac{\partial}{\partial z}\begin{bmatrix}[v_i]\\ [j_i]\end{bmatrix}+\begin{bmatrix}[0] & [L_i]\\ [C] & [0]\end{bmatrix}\frac{\partial}{\partial t}\begin{bmatrix}[v_i]\\ [j_i]\end{bmatrix}+\begin{bmatrix}[0] & [C_i]\\ [G] & [0]\end{bmatrix}\begin{bmatrix}[v_i]\\ [j_i]\end{bmatrix}=\begin{bmatrix}[0]\\ [0]\end{bmatrix} \qquad [4.76]$$

Then, we propose the electrical scheme shown in Figure 4.7 for a single lossy transmission line.

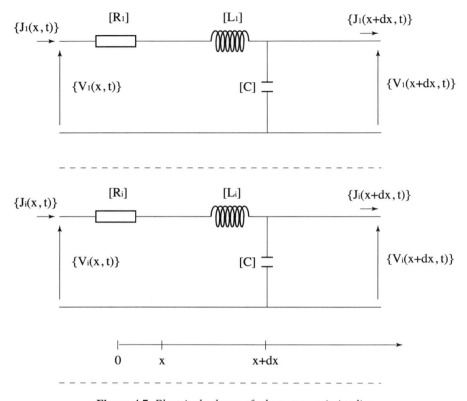

Figure 4.7. *Electrical scheme of a lossy transmission line*

4.4. Weakly bent transmission lines in the quasi-TEM approximation

4.4.1. *Bent lossy heterogenous media with lossless conductors*

The general case of *bent media of electromagnetic wave propagation* has been discussed earlier in the chapter devoted to the derivation of general integro-differential equations from Maxwell's equations. Local coordinates have been used tied to the potential vector (electric or magnetic vector). In the case of bent conductors, the vector potential is not necessarily tangential to the edge of the conductor so that the classical TM or TE modes of propagation no longer exist. Nevertheless, when the bent curvature is weak, we propose a quasi-TEM approximation according to which the axial electrical (or magnetic) field is zero. In this case, coming back to the general equation:

$$\frac{\partial J}{\partial t} = -\oint_{C_H} \frac{1}{\mu} \left[(\vec{e}\vec{\nabla}) - \frac{1}{\Re} \right] \left[E_n + (\vec{n}\vec{\nabla})V \right] dl_H + \oint_{C_H} \frac{1}{\mu} \frac{\partial}{\partial t} \left[(\vec{e}\vec{\nabla}) - \frac{1}{\Re} \right] A \, dl_H$$

we have $E_n = 0$, and therefore only the first integral of the right-hand side remains, thus giving the result

$$\frac{\partial J}{\partial t} = -\oint_{C_H} \frac{1}{\mu} \left[(\vec{e}\vec{\nabla}) - \frac{1}{\Re} \right] (\vec{n}\vec{\nabla})V \, dl_H \quad [4.77]$$

Now, on the lossless conductor boundary, we encounter the constraint $\vec{e}\vec{H} = 0$, so that the right-hand side of the equation

$$(\vec{n}\vec{\nabla})J = \oint_{C_H} \left(\sigma + \varepsilon \frac{\partial}{\partial t} \right) E_s \, dl_H - \oint_{C_H} \varepsilon \vec{E} \frac{\partial \vec{e}}{\partial t} dl_H + \oint_{C_H} \frac{\vec{e}\vec{H}}{\Re_{t_H}} dl_H$$

is reduced to its first integral, and we obtain

$$(\vec{n}\vec{\nabla})J = \oint_{C_H} \left(\sigma + \varepsilon \frac{\partial}{\partial t} \right) E_s \, dl_H \quad [4.78]$$

4.4.2. Bent lossy homogenous media with lossless conductors

Assuming the propagation media to be completely homogenous, so μ, ε and σ are constant, equations [4.77] and [4.78] become, respectively

$$\frac{\partial J}{\partial t} = -\frac{1}{\mu} \oint_{C_H} \left[(\vec{e}\vec{\nabla}) - \frac{1}{\Re} \right] (\vec{n}\vec{\nabla})V \, dl_H \quad [4.79]$$

$$(\vec{n}\vec{\nabla})J = \left(\sigma + \varepsilon \frac{\partial}{\partial t} \right) \oint_{C_H} E_s \, dl_H \quad [4.80]$$

128   Electromagnetism and Interconnections

4.4.3. Bent lossless conductors such that e_n does not depend on q_1, and e_1 and C_H do not depend on q_n

In the case of bent media such that e_n does not depend on the coordinate q_1 and both e_1 and C_H do not depend on q_n, the operators $(\vec{e}\vec{\nabla})$ and $(\vec{n}\vec{\nabla})$ are commutable, and by means of the relations $E_s = -(\vec{e}\vec{\nabla})V$ and

$$(\vec{n}\vec{\nabla})\oint_{C_H} \vec{H}\,\mathrm{d}\vec{l} = \oint_{C_H}(\vec{n}\vec{\nabla})(\vec{H}\,\mathrm{d}\vec{l})$$

equation [4.79] becomes

$$\frac{\partial J}{\partial t} = -\frac{1}{\mu}(\vec{n}\vec{\nabla})\oint_{C_H}\left[E_s + \frac{V}{\Re}\right]\mathrm{d}l_H \qquad [4.81]$$

which shows a perturbation of the electrical field equal to $V/\Re$ in the cross-section of the lossless conductor tied to the curvature of the vector potential lines assumed to travel along the conductor boundary. Furthermore, if it is known, in respect to the quasi-TEM approximation, that the conductor boundary is an equipotential surface, V then being a constant in the integral $\oint_{C_H}(V/\Re)\mathrm{d}l_H$, we obtain

$$\varepsilon\oint_{C_H}\frac{\mathrm{d}l_H}{\Re} = C_\Re$$

which defines a new *lineic capacitor tied to the curvature* of the bent media. Returning to the lineic charge Q_S, defined by [4.45], we write

$$\frac{\partial J}{\partial t} = -\frac{1}{\mu}(\vec{n}\vec{\nabla})\left[\frac{Q_s + VC_\Re}{\varepsilon}\right]$$

Equation [4.80] becomes the same as [4.43] which, then, is always valid whatever the geometry of purely conductive boundaries.

4.4.4. Lineic capacitance tied to a weak curvature of a transmission line

Here we consider the geometric features concerning the lineic capacitor as depicted in Figure 4.8 ($\vec{n}$ oriented straight section).

Transmission Line Equations 129

Let (C_H) be the closed curve ABCD of the lossless conductor boundary. We apply the Chasles relation as follows:

$$\oint_{C_H} \frac{dl_H}{\Re} = \int_{AB} \frac{dl_H}{\Re} + \int_{BC} \frac{dl_H}{\Re} + \int_{CD} \frac{dl_H}{\Re} + \int_{DA} \frac{dl_H}{\Re}$$

$$= \int_{AB} d\varphi + \int_{\Re_1(\omega_1)}^{\Re_2(\omega_2)} \frac{d\Re}{\Re} + \int_{CD} d\varphi + \int_{\Re_1(0)}^{\Re_2(0)} \frac{d\Re}{\Re}$$

$$= \delta\varphi_1 - \delta\varphi_2 + \ln\frac{\Re_2(\omega_2)\Re_1(0)}{\Re_1(\omega_1)\Re_2(0)}$$

From this, we obtain the capacitance per unit length tied to the curvature of the conductors:

$$C_\Re = \varepsilon\left[\delta\varphi_1 - \delta\varphi_2 + \ln\frac{\Re_2(\omega_2)\Re_1(0)}{\Re_1(\omega_1)\Re_2(0)}\right]$$

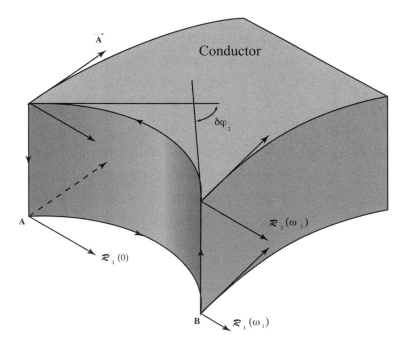

Figure 4.8. *Bent conductor geometry*

4.5. Conclusion

The integro-differential equations of electrical (respectively magnetic) current obtained in the previous chapter become transmission line equations with second term depending on electrical (respectively magnetic) longitudinal fields. This second term is not zero in the general case of any lossy homogenous interconnection media traveled by any propagation mode waves. Its being zero is met by the classical TEM propagation mode of any signal having a wavelength range beyond the geometrical dimensions of the technological structures where it is traveling.

Furthermore the previous theory of electromagnetism in straight generalized stratified media have shown the way for a Foucault current spectral analysis in the case of losses inside even coupled conductors.

Chapter 5

Direct Time-domain Methods

Three time-domain methods of solving the different equations of propagation along interconnections are presented in this book.

This chapter deals with the first method, called *direct*, and is strictly rigorous and devoted only to lossless propagation media. After a classical review of solving the lossless propagation equation, so defining the forward and backward waves, an original approach to solving the matrix transmission line equation, by means of the spectral theory of matrix functions depending on several commutable matrix variables, is proposed, accounting directly for the initial and boundary conditions. This leads us to define the *propagator* or linear operator of lossless wave propagation, so dealing easily with the multiple reflection analysis along homogenous and heterogenous lines as well. This development is generalized to any set of lossless coupled transmission lines having the same crosstalk length. Then, the bifurcations are accounted for at their connecting nodes with respect to Kirchoff's laws, so any complete complex network modeling inside any multilayered substrate occurs by chaining down all the line segment matrix equations into only one called the *matrix state equation* of the electronic system. To conclude, computation and memory charges are estimated.

5.1. "Direct" methods in the time domain

5.1.1. *Defining a "direct" method in the time domain*

A method of solving transmission line or propagation equations is called "direct" if the rigorous time-domain solution (without any approximation) depends only on the present time and on a finite number of times in the past.

5.1.2. *Single lossless transmission lines in homogenous media*

5.1.2.1. *Forward waves, backward waves*

The lossless telegrapher's equation corresponds to those (equations [4.49], [4.50], [4.54], and [4.55]) presented in Chapter 4 when $\sigma = 0$, so that we obtain the following classical form which is called the propagation equation:

$$\frac{\partial^2 u}{\partial z^2} - \frac{1}{v^2}\frac{\partial^2 u}{\partial t^2} = 0 \qquad [5.1]$$

where u represents any one of the features J, φ, A, or V encountered before, and $v = 1/\sqrt{\mu\varepsilon}$ represents the wave edge velocity which is assumed to be constant here. Let us review briefly how to solve efficiently the propagation equation by means of an operational calculus and of classical variables transform which we would like to extend to the case of the lossy line analysis later on directly in the time domain without the need for any functional transform like the Fourier or Laplace transforms.

Indeed, we can write

$$\left(\frac{\partial}{\partial z} + \frac{1}{v}\frac{\partial}{\partial t}\right)\left(\frac{\partial}{\partial z} - \frac{1}{v}\frac{\partial}{\partial t}\right)u = 0 \qquad [5.2]$$

Let the new variables be

$$\begin{cases} t = \xi + \xi' \\ z = v(\xi - \xi') \end{cases}$$

We now have

$$\frac{\partial}{\partial z} + \frac{1}{v}\frac{\partial}{\partial t} = \frac{1}{v}\frac{\partial}{\partial \xi}$$

$$\frac{\partial}{\partial z} - \frac{1}{v}\frac{\partial}{\partial t} = -\frac{1}{v}\frac{\partial}{\partial \xi'}$$

so that [5.2] becomes

$$\frac{\partial^2 u}{\partial \xi \partial \xi'} = 0$$

After two integrations with respect to these new variables and returning to the initial ones z and t, we obtain the following very classical result:

$$u(z,t) = f\left(t - \frac{z}{v}\right) + g\left(t + \frac{z}{v}\right) \quad [5.3]$$

f and g being two twice continuously differentiable functions the physical meaning of which is obvious. The function f depending on $(t - z/v)$ is called *forward wave*; g depending on $(t + z/v)$ is called the *backward wave*.

Indeed, we have to know what functions f and g are according to the initial conditions or versus time at a given point z_0. At this step of our analysis, no reflecting obstacle is considered: the analysis concerns the propagation of so-called "free waves".

Let us note that the waveform does not change during propagation if and only if either f or g is zero: in this case, we talk about the so-called "isomorphism of the waveform shape".

In any case, it remains to solve practically the problem satisfying both the first-order transmission line equations:

$$\begin{cases} \dfrac{\partial J}{\partial t} = -\dfrac{1}{\mu\varepsilon}\dfrac{\partial Q}{\partial z} \\ \dfrac{\partial J}{\partial z} = -\dfrac{\partial Q}{\partial t} \end{cases}$$

or

$$\begin{cases} \dfrac{\partial A}{\partial z} + \mu\varepsilon \dfrac{\partial V}{\partial t} = 0 \\ \dfrac{\partial A}{\partial t} + \dfrac{\partial V}{\partial z} = 0 \end{cases}$$

with the initial and/or boundary conditions when we are solving the telegrapher's equations:

$$\begin{cases} \dfrac{\partial^2 J}{\partial z^2} - \mu\varepsilon \dfrac{\partial^2 J}{\partial t^2} = 0 \\ \dfrac{\partial^2 Q}{\partial z^2} - \mu\varepsilon \dfrac{\partial^2 Q}{\partial t^2} = 0 \end{cases}$$

or

$$\begin{cases} \dfrac{\partial^2 A}{\partial z^2} - \mu\varepsilon \dfrac{\partial^2 A}{\partial t^2} = 0 \\ \dfrac{\partial^2 V}{\partial z^2} - \mu\varepsilon \dfrac{\partial^2 V}{\partial t^2} = 0 \end{cases}$$

Their solutions consist of four unknown functions that have to be computed with respect to the above mentioned constrains: first-order partial differential equations and the initial or boundary conditions. This cannot be achieved easily by directly using the forward and backward waves coming from second-order equations.

5.1.2.2. Propagators group of the homogenous single transmission line equation

We propose a new approach to solving directly the first-order transmission line equations by writing them in matrix form and by considering them like a first-order partial differential equation with respect to commutable matrix variables. So, the solution is obtained as a matrix function of these matrix variables for which the use of the so-called Duncan-Sylvester spectral expansion formula directly gives a physically meaningful expression showing the expected forward and backward waves corresponding to the initial or boundary conditions.

The transmission line equations can be written as follows:

$$\dfrac{\partial}{\partial t}\begin{bmatrix} Q \\ J \end{bmatrix} + \begin{bmatrix} 0 & 1 \\ v^2 & 0 \end{bmatrix} \dfrac{\partial}{\partial z}\begin{bmatrix} Q \\ J \end{bmatrix} = \begin{bmatrix} 0 \\ 0 \end{bmatrix} \qquad [5.4]$$

where $v = 1/\sqrt{\mu\varepsilon}$.

Let us demonstrate the method of our matrix computation. We define the following commutable matrix variables:

$$[\tau] = t\begin{bmatrix} 1 & 0 \\ 0 & 1 \end{bmatrix} \quad \text{and} \quad [\zeta] = z\begin{bmatrix} 0 & 1 \\ v^2 & 0 \end{bmatrix}^{-1} = z\begin{bmatrix} 0 & \dfrac{1}{v^2} \\ 1 & 0 \end{bmatrix}$$

Then, the matrix transmission line equation becomes

$$\left\{\frac{\partial}{\partial[\zeta]} + \frac{\partial}{\partial[\tau]}\right\}[\varXi]\begin{bmatrix} Q_0 \\ J_0 \end{bmatrix} = \begin{bmatrix} 0 \\ 0 \end{bmatrix}$$

The method is to find the solution as a matrix function of the two above commutable matrix variables $[\zeta]$ and $[\tau]$ as

$$\begin{bmatrix} Q(z,t) \\ J(z,t) \end{bmatrix} = [\varXi([\zeta],[\tau])]\begin{bmatrix} Q_0 \\ J_0 \end{bmatrix}$$

J_0 and Q_0 being constant scalars. Following the case where the variables are scalars, the solution is found as a once continuously differential matrix function of the matrix variable:

$$[\xi] = [\tau] - [\zeta] = \begin{bmatrix} t & -\dfrac{z}{v^2} \\ -z & t \end{bmatrix} \qquad [5.5]$$

We now use the Duncan–Sylvester spectral expansion formula that we recall here. $[A]$ is a second-order matrix having the different eigenvalues λ_1 and λ_2, and we have the formula

$$f(A) = f(\lambda_1)\frac{[A] - \lambda_2[1]}{\lambda_1 - \lambda_2} + f(\lambda_2)\frac{[A] - \lambda_1[1]}{\lambda_2 - \lambda_1}$$

Here, $[A]$ is equal to $[\xi]$, and the eigenvalues of the matrix $[\xi]$ are solutions of the characteristic equation

$$\det\{\xi - \lambda[1]\} = \begin{vmatrix} t-\lambda & -\dfrac{z}{v^2} \\ -z & t-\lambda \end{vmatrix} = (t-\lambda)^2 - \dfrac{z^2}{v^2}$$

Therefore, they are $\lambda_1 = t - z/v$ and $\lambda_2 = t + z/v$, and we obtain

$$[\Xi(z,t)] = f([\xi]) = \dfrac{1}{2}\left(t - \dfrac{z}{v}\right)\begin{bmatrix} 1 & \dfrac{1}{v} \\ v & 1 \end{bmatrix} + \dfrac{1}{2}f\left(t + \dfrac{z}{v}\right)\begin{bmatrix} 1 & -\dfrac{1}{v} \\ -v & 1 \end{bmatrix} \qquad [5.6]$$

Assuming z is positive, this shows the forward wave at the first term of the right-hand side and the backward wave at the second term. If z is negative, the previous terms are exchanged. In both cases, v is the velocity of the wave edge. The time $t = z/v$ is the propagation delay of the wave edge from the point $z = 0$ to the point z.

Let f be a function so that $f(u < 0) = 0$. Either we know the initial conditions, which are similar for both current and lineic charge in this case:

$$[\Xi(z,0)] = \dfrac{1}{2}f\left(-\dfrac{z}{v}\right)\begin{bmatrix} 1 & \dfrac{1}{v} \\ v & 1 \end{bmatrix} + \dfrac{1}{2}f\left(\dfrac{z}{v}\right)\begin{bmatrix} 1 & -\dfrac{1}{v} \\ -v & 1 \end{bmatrix}$$

and we can know the evolution of the wave at every time later, or we know the boundary conditions, again similar for both current and lineic charge, at the point $z = 0$:

$$[\Xi(0,t)] = f(t)\begin{bmatrix} 1 & 0 \\ 0 & 1 \end{bmatrix}$$

We define, in this last case, the *propagator* as the linear operator $[P(z)]$ as follows:

$$[\Xi(0,t)] \xrightarrow{[P(z)]} [\Xi(z,t)] = [P(z)][\Xi(0,t)] \qquad [5.7]$$

Using the translation operator τ_a so that $f(t) \xrightarrow{\tau_a} f(t-a)$, we can write

$$[P(z)] = \begin{bmatrix} \frac{1}{2} & \frac{1}{2v} \\ \frac{v}{2} & \frac{1}{2} \end{bmatrix} \tau_{z/v} + \begin{bmatrix} \frac{1}{2} & -\frac{1}{2v} \\ -\frac{v}{2} & \frac{1}{2} \end{bmatrix} \tau_{-z/v} \qquad [5.8]$$

This shows the *spectral expansion* of the matrix propagator. From the theory of matrix algebra, the matrices involved on the right-hand side are called "projectors", having the properties that their square is equal to themselves and the product of two different ones is zero.

These properties lead to the relations

$$\left.\begin{array}{l}[P(z+z')] = [P(z)][P(z')] \\ [P(0)] = [1] \end{array}\right\} \qquad [5.9]$$

that give the set of propagators a *group structure* which is useful for processing several transmission lines in series as required in applications (*propagator group*).

Nevertheless, in real applications, we need to process more general situations like those where the current $J_0(t)$ and the lineic charge $Q_0(t)$ are not similar. In that case, we have to write

$$\left.\begin{array}{l} \begin{bmatrix} Q_0(t) \\ J_0(t) \end{bmatrix} = [\Xi(0,t)] \begin{bmatrix} 1 \\ 0 \end{bmatrix} + [\Xi'(0,t)] \begin{bmatrix} 0 \\ 1 \end{bmatrix} \\ \begin{bmatrix} Q(z,t) \\ J(z,t) \end{bmatrix} = [P(z)] \begin{bmatrix} Q_0(t) \\ J_0(t) \end{bmatrix} \end{array}\right\} \qquad [5.10]$$

So we are able to take account of boundary conditions at one point. However, we have to be aware that boundary conditions are needed at two points in real applications: at one point the current is involved, at the other the lineic charge is involved or, in the quasi-TEM assumption, the voltage.

5.1.2.3. *Incident and reflected waves, characteristic impedance, moving and standing waves*

Let us deal with the practical case of boundary conditions at two different points, so meeting the classical analysis devoted to stand-alone lossless transmission lines directly in the time domain that we want to extend to complex networks of lossy coupled lines. Assuming the quasi-TEM mode of propagation, the lineic charge Q is related to the voltage V which is well defined in this case by means of the relation

$Q = CV$ applicable everywhere along the transmission line. Then the solution of the propagation equation becomes

$$\begin{bmatrix} V(z,t) \\ J(z,t) \end{bmatrix} = \begin{bmatrix} \dfrac{1}{C} & 0 \\ 0 & 1 \end{bmatrix} [P(z)] \begin{bmatrix} C & 0 \\ 0 & 1 \end{bmatrix} \begin{bmatrix} V_0(t) \\ J_0(t) \end{bmatrix}$$

thus introducing the operational transfer matrix:

$$\begin{bmatrix} V(z,t) \\ J(z,t) \end{bmatrix} = \underbrace{\begin{bmatrix} \dfrac{1}{2}(\tau_{z/v} + \tau_{-z/v}) & \dfrac{1}{2vC}(\tau_{z/v} - \tau_{-z/v}) \\ \dfrac{vC}{2}(\tau_{z/v} - \tau_{-z/v}) & \dfrac{1}{2}(\tau_{z/v} + \tau_{-z/v}) \end{bmatrix}}_{\text{operational transfer matrix}} \begin{bmatrix} V_0(t) \\ J_0(t) \end{bmatrix} \qquad [5.11]$$

Let R be a *resistive load* at the line terminal, so that $V = RJ$. By changing z to $(-z)$, by means of the above mentioned properties [5.9] of the propagator, we obtain from the previous equation where $V(z, t) = f(t)$

$$V_0(t) = \underbrace{\frac{1}{2}\left(1 + \frac{1}{RvC}\right) f\left(t + \frac{z}{v}\right)}_{\text{incident wave}} + \underbrace{\frac{1}{2}\left(1 - \frac{1}{RvC}\right) f\left(t - \frac{z}{v}\right)}_{\text{reflected wave}} \qquad [5.12]$$

thus defining the incident and the reflected waves.

The reflection ratio is defined as

$$\Re = \frac{1 - \dfrac{1}{RvC}}{1 + \dfrac{1}{RvC}} \qquad [5.13]$$

If $1/vC = R$, there is no reflection. Then the quantity $1/vC$, having the same dimension as a resistor, is called the *characteristic impedance* Z_{c0} of the lossless transmission line.

The reflection ratio being zero, the wave expression is reduced to $f(t + z/v)$ which defines a *moving wave*.

In contrast, when the reflection ratio is equal to 1, there is an open circuit at the line terminal, and the wave expression becomes

$$\frac{1}{2}\left[f\left(t-\frac{z}{v}\right)+f\left(t+\frac{z}{v}\right)\right]$$

which defines a standing wave.

In the general case of any wave having its reflection ratio between 0 and 1, this wave can be expanded into the sum of a moving wave and a standing wave as follows:

$$V_0(t) = \underbrace{\frac{Z_{c0}}{R}f\left(t+\frac{z}{v}\right)}_{\text{moving wave}} + \underbrace{\left(1-\frac{Z_{c0}}{R}\right)\frac{1}{2}\left[f\left(t-\frac{z}{v}\right)+f\left(t+\frac{z}{v}\right)\right]}_{\text{standing wave}} \qquad [5.14]$$

5.1.2.4. *Multiple reflection processing*

Using the characteristic impedance $Z_{c0} = 1/VC$, the reflection ratio obtains its very classical form:

$$\Re = \frac{R-Z_{c0}}{R+Z_{c0}} \qquad [5.15]$$

Processing the time domain translation $t \to t - z/v$ in the wave expression, setting the past time terms assumed to be known on the right-hand side and the unknown present time term on the left-hand side we get the recursive relation leading to the multiple reflections:

$$f(t) = \underbrace{(1-\Re)V_0\left(t-\frac{z}{v}\right)}_{\text{forward wave}} - \underbrace{\Re f\left(t-\frac{2z}{v}\right)}_{\text{backward wave}} \qquad [5.16]$$

Let $f(u) = 0$; when $u < 0$, $V(t)$ takes off since $t = z/v$, which is the traveling time to the line terminal. A first reflection occurs at the time $t = 2z/v$, which is the forward and back delay.

After this, the wave begins to be fed with the reflected wave

$$\Re f\left(t-\frac{2z}{v}\right)$$

during $0 \le t - 2z/v \le 2z/v$, which gives

140   Electromagnetism and Interconnections

$$(1-\Re)V_0\left(t-\frac{3z}{v}\right)$$

during $3z/v < t < 4z$.

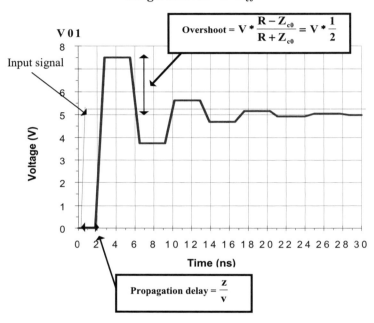

Figure 5.1. *Single line with $R > Z_{c0}$*

So step by step, the multiple reflected wave can be formulated as shown in Figures 5.1 and 5.2, first with $R > Z_{c0}$ then with $R < Z_{c0}$.

Direct Time-domain Methods   141

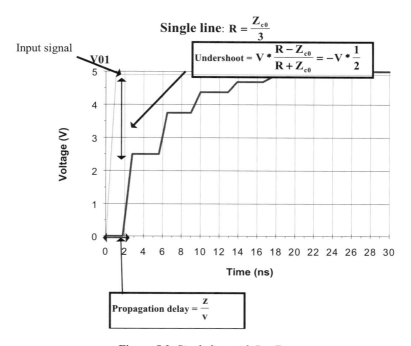

Figure 5.2. *Single line with $R < Z_{c0}$*

Let us now deal with a heterogenous single transmission line made up of several homogenous single transmission lines in series, as shown in Figure 5.3.

These transmission lines generate a very complex situation, in respect of the multiple reflections created at each interface between two adjacent segments, which can be depicted by the diagram shown in Figure 5.4, where the time flows from the top to the bottom and the distances are counted from the left to the right.

This diagram shows all the available paths of the reflected pulses traveling along the heterogenous line at any time. Going down along the right diagonal, there are the set of pulses which were reflected an even number times, the number of double reflections being called the *multiplicity order*. Going down along the left diagonal gives the set of pulses having met an odd number of reflections.

142   Electromagnetism and Interconnections

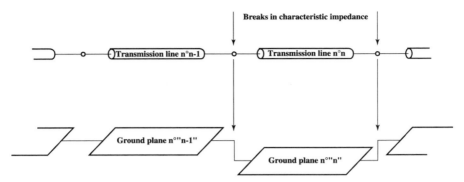

Figure 5.3. *Heterogenous transmission line*

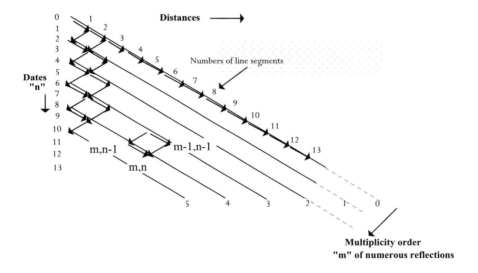

Figure 5.4. *Numerous reflections diagram along a heterogenous transmission line (m = number of double reflections needed for meeting a given point; n = number of time steps or segments traveled without reflection or the maximum number of reflections met on the first segment at a given time)*

The number of paths available for any pulse having met m double reflections at the time n is equal to the number of combinations $P_{m,n} = C_n^m$ because Figure 5.4 shows the relations $P_{m,n} = P_{m,n-1} + P_{m-1,n-1}$, $P_{0,n} = 1$. The number of incident and reflected pulses traveling at the time n is equal to $C_n^0 + C_n^1 + \cdots + C_n^{m'}$, m' being the integer closest to $n/2$ which is rigorously $2n - 1$ when n is even.

Direct Time-domain Methods   143

5.2. Lossless coupled transmission lines in homogenous media

5.2.1. *Homogenous coupling*

Let us assume that the lossless coupled transmission lines have the same length and their terminals are set within the same cross-section. As the propagation mode is quasi-TEM, every electromagnetic wave has the same velocity v, and it is possible to define, in this case, capacitance and inductance matrices per unit length $[C]$ and $[L]$, respectively, to which the characteristic impedance

$$[Z_{c0}] = \frac{1}{v}[C]^{-1} = v[L] \qquad [5.17]$$

is related (numerical computation at the end of this chapter).

Then, the *electromagnetic influence matrix* between the current and voltage vectors is defined as the matrix block

$$v\begin{bmatrix}[0] & [L]\\ [C] & [0]\end{bmatrix} = [M] = \begin{bmatrix}[0] & [Z_{c0}]\\ [Z_{c0}]^{-1} & [0]\end{bmatrix} \qquad [5.18]$$

so that the coupled lossless transmission line equations [4.68] and [4.69] can be written as

$$\frac{\partial}{\partial t}\begin{bmatrix}[V]\\ [J]\end{bmatrix} + v[M]^{-1}\frac{\partial}{\partial z}\begin{bmatrix}[V]\\ [J]\end{bmatrix} = \begin{bmatrix}[0]\\ [0]\end{bmatrix} \qquad [5.19]$$

Since the matrix $[M]$ is involutive (its square is equal to the unit matrix), it has only two different eigenvalues (-1) and $(+1)$ which have a multiplicity order at least equal to 2 since its dimension is greater than 2×2. In this case, the Duncan–Sylvester approach is no longer applicable as previously shown for the stand-alone lines. We have to proceed as follows.

Using the relation $[V] = [C]^{-1}[Q] = v[Z_{c0}][Q]$, equation [5.19] becomes (because $[M]^{-1} = [M]$)

$$\begin{bmatrix}v[Z_{c0}] & [0]\\ [0] & [1]\end{bmatrix}\frac{\partial}{\partial t}\begin{bmatrix}[Q]\\ [J]\end{bmatrix} + v\begin{bmatrix}[0] & [Z_{c0}]\\ [Z_{c0}]^{-1} & [0]\end{bmatrix}\begin{bmatrix}v[Z_{c0}] & [0]\\ [0] & [1]\end{bmatrix}\frac{\partial}{\partial z}\begin{bmatrix}[Q]\\ [J]\end{bmatrix} = \begin{bmatrix}[0]\\ [0]\end{bmatrix}$$

Multiplying on the left-hand side by

$$\begin{bmatrix} \frac{1}{v}[Z_{c0}]^{-1} & [0] \\ [0] & [1] \end{bmatrix}$$

gives

$$\frac{\partial}{\partial t}\begin{bmatrix} [Q] \\ [J] \end{bmatrix} + \begin{bmatrix} [0] & [1] \\ v^2[1] & [0] \end{bmatrix}\frac{\partial}{\partial z}\begin{bmatrix} [Q] \\ [J] \end{bmatrix} = \begin{bmatrix} [0] \\ [0] \end{bmatrix}$$

There are as many single line equations [5.4] as there are coupled lines.

Therefore, we apply directly result [5.10]:

$$\begin{bmatrix} [Q] \\ [J] \end{bmatrix} = [P(z)]\begin{bmatrix} [Q_0] \\ [J_0] \end{bmatrix}$$

with the following propagator $[P(z)]$ made up of sub-matrices:

$$[P(z)] = \begin{bmatrix} \frac{1}{2}[1] & \frac{1}{2v}[1] \\ \frac{v}{2}[1] & \frac{1}{2}[1] \end{bmatrix}\tau_{z/v} + \begin{bmatrix} \frac{1}{2}[1] & -\frac{1}{2v}[1] \\ -\frac{v}{2}[1] & \frac{1}{2}[1] \end{bmatrix}\tau_{-z/v}$$

Then, we obtain the transfer operational relation

$$\begin{bmatrix} [V] \\ [J] \end{bmatrix} = [T]\begin{bmatrix} [V_0] \\ [J_0] \end{bmatrix}$$

$[T]$ being the transfer operational matrix. By means of the previous propagator and of the relation $[V] = v[Z_{c0}][Q]$, we obtain

$$[T] = \frac{1}{2}\begin{bmatrix} [1] & [Z_{c0}] \\ [Z_{c0}]^{-1} & [1] \end{bmatrix}\tau_{z/v} + \frac{1}{2}\begin{bmatrix} [1] & -[Z_{c0}] \\ -[Z_{c0}]^{-1} & [1] \end{bmatrix}\tau_{-z/v}$$

Note that $[T(z)]^{-1} = [T(-z)]$, so giving

$$\begin{bmatrix} [V_0] \\ [J_0] \end{bmatrix} = [T(-z)]\begin{bmatrix} [V] \\ [J] \end{bmatrix} \qquad [5.20]$$

Direct Time-domain Methods 145

from which we obtain

$$[V_0(t)] = \frac{1}{2}\left[V\left(t+\frac{z}{v}\right)\right] + \frac{[Z_{c0}]}{2}\left[J\left(t+\frac{z}{v}\right)\right] + \frac{1}{2}\left[V\left(t-\frac{z}{v}\right)\right] - \frac{[Z_{c0}]}{2}\left[J\left(t-\frac{z}{v}\right)\right]$$

$$[J_0(t)] = \frac{[Z_{c0}]^{-1}}{2}\left[V\left(t+\frac{z}{v}\right)\right] + \frac{1}{2}\left[J\left(t+\frac{z}{v}\right)\right] - \frac{[Z_{c0}]^{-1}}{2}\left[V\left(t-\frac{z}{v}\right)\right] + \frac{1}{2}\left[J\left(t-\frac{z}{v}\right)\right]$$

Now we load the coupled lines by way of a purely resistive network characterized by the matrix $[R]$ so that

$$[V] = [R][J]$$

Equation [5.26] is processed by means of the time shift $t \to t - z/v$.

We get, from the above equation, the following results:

$$\left[V_0\left(t-\frac{z}{v}\right)\right] = \frac{1}{2}\{[1]+[Z_{c0}][R]^{-1}\}[V(t)] + \frac{1}{2}\{[1]-[Z_{c0}][R]^{-1}\}\left[V\left(t-\frac{2z}{v}\right)\right]$$

[5.21]

$$\left[J_0\left(t-\frac{z}{v}\right)\right] = \frac{1}{2}\{[R]^{-1}+[Z_{c0}]^{-1}\}[V(t)] + \frac{1}{2}\{[R]^{-1}-[Z_{c0}]^{-1}\}\left[V\left(t-\frac{2z}{v}\right)\right]$$

[5.22]

Rigorously, any impedance matching of all the coupled lines set together requires that

$$[R] = [Z_{c0}] = \frac{1}{v}[C]^{-1}$$

The corresponding electrical scheme is shown in Figure 5.5.

According to the computation made for the stand-alone lines which used intensively the properties of the propagator, we obtain the relation leading the multiple reflections into the coupled lossless transmission lines:

146   Electromagnetism and Interconnections

$$[V(t)] = \underbrace{\{1-[\Re]\}\left[V_0\left(t-\frac{z}{v}\right)\right]}_{\text{transmitted wave}} - \underbrace{[\Re]\left[V\left(t-\frac{2z}{v}\right)\right]}_{\text{reflected wave}} \qquad [5.23]$$

where $[\Re] = \{1+[Z_{c0}][R]^{-1}\}^{-1}\{1-[Z_{c0}][R]^{-1}\}$ is the *scattering matrix* related to the lossless coupled lines.

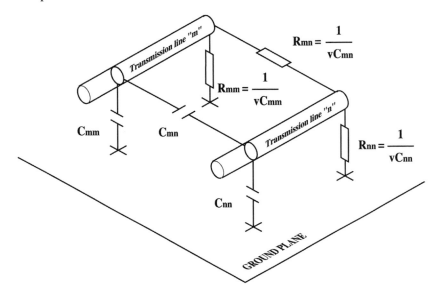

Figure 5.5. *Impedance matching of coupled lossless lines*

Figure 5.6 shows an example of crosstalk between two lines, the first one having an input voltage source, the second one being the perturbed line.

The following are the assumptions:

— both lines are coplanar and have the same geometrical features, therefore $Z_{c11} = Z_{c22}$, and $Z_{c12} = Z_{c21}$;
— these lines have the same length in crosstalk: $\tau_1 = \tau'_1 = \tau$;
— line 1 input is an impedance-free voltage source: $Z_{out} = 0$;
— line 1 has the output load $R_{in1} = Z_{c11}$;
— line 2 has equal input and output loads: $R'_{out} = R'_{in} = Z_{c11}$.

Direct Time-domain Methods   147

The following are the equations:

Line 1
$$\begin{cases} V_1(t-\tau)+\dfrac{Z_{c11}}{2}i_0(t)+\dfrac{Z_{c12}}{2}i'_0(t)=\dfrac{v_0(t)}{2}+\dfrac{v_0(t-2\tau)}{2}+\dfrac{Z_{c11}}{2}i_0(t-2\tau)+\dfrac{Z_{c12}}{2}i'_0(t-2\tau) \\ I_1(t-\tau)-\dfrac{i_0(t)}{2}-\dfrac{Z_{c12}}{2\Delta}v'(t)=-\dfrac{Z_{c11}}{2\Delta}v_0(t)+\dfrac{Z_{c11}}{2\Delta}v_0(t-2\tau)+\dfrac{i_0(t-2\tau)}{2}-\dfrac{Z_{c12}}{2\Delta}v'_0(t-2\tau) \\ v_1(t-\tau)=Z_{c11}I_1(t-\tau) \\ V_1(t-\tau)=v_1(t-\tau) \end{cases}$$

where $\Delta = (Z_{c11})^2 - (Z_{c12})^2$, and

Line 2
$$\begin{cases} V'_1(t-\tau)+\dfrac{Z_{c11}}{2}i'_0(t)+\dfrac{Z_{c12}}{2}i_0(t)-\dfrac{v'_0(t)}{2}=\dfrac{v'_0(t-2\tau)}{2}+\dfrac{Z_{c11}}{2}i'_0(t-2\tau)+\dfrac{Z_{c12}}{2}i_0(t-2\tau) \\ I'_1(t-\tau)-\dfrac{i'_0(t)}{2}-\dfrac{Z_{c12}}{2\Delta}v'_0(t)=\dfrac{Z_{c12}}{2\Delta}v_0(t)+\dfrac{Z_{c11}}{2\Delta}v'_0(t-2\tau)+\dfrac{i'_0(t-2\tau)}{2}-\dfrac{Z_{c12}}{2\Delta}v_0(t-2\tau) \\ v'_0(t)=-Z_{c11}i'_0(t) \\ v'_1(t-\tau)=Z_{c11}I'_1(t-\tau) \\ V'_1(t-\tau)=v'_1(t-\tau) \end{cases}$$

where $\Delta = (Z_{c11})^2 - (Z_{c12})^2$.

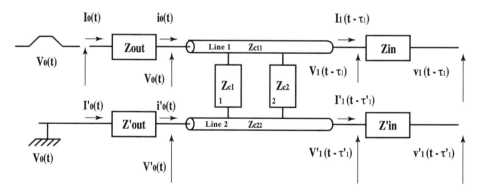

Figure 5.6. *Example of crosstalk*

The following are the formulae of waveforms for the first segments of time:

$$t < 2\tau \quad i_0(t) = \frac{2Z_{c11}}{2Z_{c11}^2 - Z_{c12}^2} v_0(t)$$

$$v_0'(t) = \frac{Z_{c11}Z_{c12}}{2Z_{c11}^2 - Z_{c12}^2} v_0(t)$$

$$2\tau < t < 4\tau \quad i_0(t) = \frac{2Z_{c11}}{2Z_{c11}^2 - Z_{c12}^2} v_0(t) - \frac{2Z_{c11}Z_{c12}^2(4Z_{c11}^2 - 5Z_{c12}^2)}{(2Z_{c11}^2 - Z_{c12}^2)(4Z_{c11}^2 - Z_{c12}^2)} v_0(t - 2\tau)$$

$$v_0'(t) = \frac{Z_{c11}Z_{c12}}{2Z_{c11}^2 - Z_{c12}^2} v_0(t) - \frac{2Z_{c11}Z_{c12}(Z_{c11}^2 - Z_{c12}^2)(4Z_{c11}^2 + Z_{c12}^2)}{(2Z_{c11}^2 - Z_{c12}^2)(4Z_{c11}^2 - Z_{c12}^2)} v_0(t - 2\tau)$$

$$v_1(t - \tau) = \frac{2Z_{c11}^2(4Z_{c11}^2 - 3Z_{c12}^2)}{(2Z_{c11}^2 - Z_{c12}^2)(4Z_{c11}^2 - Z_{c12}^2)} v_0(t - 2\tau)$$

$$v_1'(t - \tau) = \frac{-4Z_{c11}Z_{c12}^3(Z_{c11}^2 - 2Z_{c12}^2)}{(2Z_{c11}^2 - Z_{c12}^2)^2(4Z_{c11}^2 - Z_{c12}^2)} v_0(t - 2\tau)$$

Figures 5.7–5.10 depict the voltages at the terminal of both active and perturbed lines according to the above equations in the two major cases of weak coupling where the disturbed line does not have any significant influence on the active line and of strong coupling where there are mutual influences between the two lines.

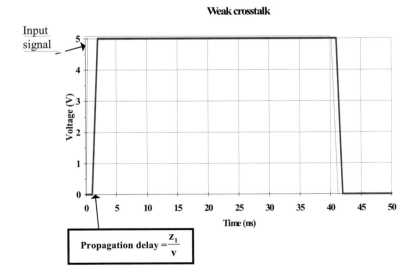

Figure 5.7. *Active line in the case of weak coupling*

Direct Time-domain Methods 149

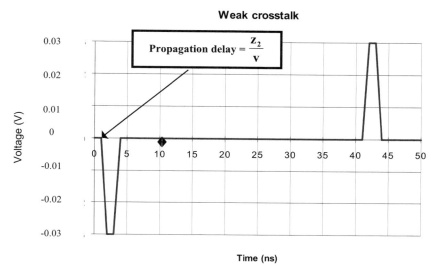

Figure 5.8. *Perturbed line in the case of weak coupling*

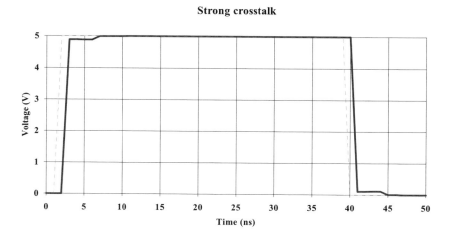

Figure 5.9. *Active line perturbed in the case of strong coupling*

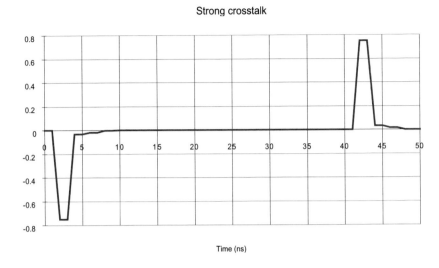

Figure 5.10. *Perturbed line in the case of strong coupling*

5.2.2. *Heterogenous coupling*

Let us start with heterogenously coupled segments, as depicted in Figure 5.11. This case is processed by way of the matrix blocks as follows:

$$\begin{bmatrix} \left[V_i'(z_1+z_2',t-\dfrac{z_2'}{v})\right] \\ \left[J_i'(z_1+z_2',t-\dfrac{z_2'}{v})\right] \\ \left[V_i''(z_1+z_2'',t-\dfrac{z_2''}{v})\right] \\ \left[J_i''(z_1+z_2'',t-\dfrac{z_2''}{v})\right] \end{bmatrix} - \begin{bmatrix} [T_i'] & [0] \\ [0] & [T_i''] \end{bmatrix} \begin{bmatrix} V_i'(z_1,t) \\ J_i'(z_1,t) \\ V_i''(z_1,t) \\ J_i''(z_1,t) \end{bmatrix} = \begin{bmatrix} [T_i'] & [0] \\ [0] & [T_i''] \end{bmatrix} \begin{bmatrix} \left[V_i'(z_1,t-\dfrac{2z_2'}{v})\right] \\ \left[J_i'(z_1,t-\dfrac{2z_2'}{v})\right] \\ \left[V_i''(z_1,t-\dfrac{2z_2''}{v})\right] \\ \left[J_i''(z_1,t-\dfrac{2z_2''}{v})\right] \end{bmatrix}$$

[5.24]

where the matrices $[T_i']$ and $[T_i'']$ have the form

$$\frac{1}{2}\begin{bmatrix} [1] & [Z_{c0}] \\ [Z_{c0}]^{-1} & [1] \end{bmatrix}$$

in which $[Z_{c0}]$ is the characteristic impedance of each set of coupled lines. This matrix block procedure is extended to lumped networks at the coupled lines terminal.

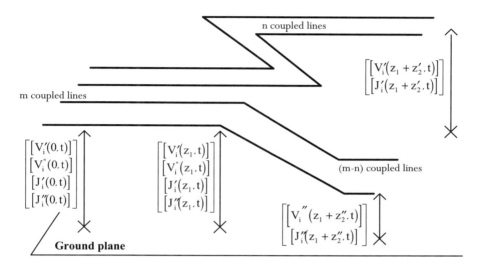

Figure 5.11. *Heterogenous coupling of lossless transmission lines*

5.2.3. Bifurcations

Let us now deal with the bifurcations, as shown in Figure 5.12. The classical Kirchoff's laws are applicable at each bifurcation node: the same voltage occurs for every branch of the bifurcation, while the sum of all the currents arriving at the node is zero. This means that the following equations have to be applied:

$$\left. \begin{array}{l} [V_i^{(k)}(z,t)] - [V_i^{(1)}(z,t)] = [0] \\ \sum_k [J_i^{(k)}(z,t)] = [0] \end{array} \right\} \qquad [5.25]$$

We deal with a single line made up of a single bifurcation then three segments having a common point.

Figure 5.13 shows the equivalent electrical scheme of this line with the models of its terminal nodes.

152   Electromagnetism and Interconnections

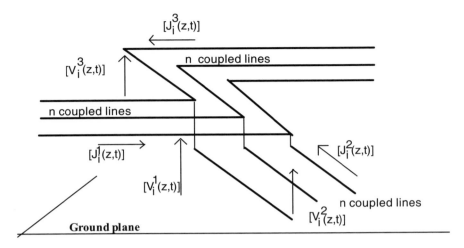

Figure 5.12. *Bifurcations between coupled lossless transmission lines*

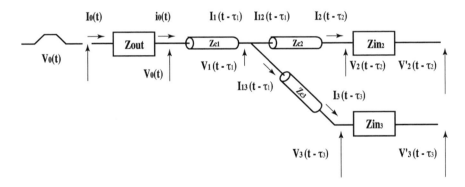

Figure 5.13. *Example of bifurcation*

The assumptions are as follows. The three segments have the same *l*, to which the propagation delays are related:

$$\tau_1 = \tau_2 = \tau_3 = \tau = \frac{l\sqrt{\varepsilon_r}}{c}$$

where c is the velocity of light in a vacuum ($= 3 \times 10^8$ m s^{-1}) and ε_r is a constant dielectric parameter of the propagation media.

Segments 2 and 3 are set on the same layer of the printed circuit board so that

$$Z_{c1} = 2Z_{c2} = 2Z_{c3} = 2Z_c$$

and they are loaded by the impedances $Z_{in2} = Z_{in3} = Z_c$.

At the line input, $Z_{out} = 0$ which creates reflections of any reflected wave coming back from the bifurcation.

Therefore we obtain the equations where the unknown currents and voltages are set on the left-hand side:

1st segment
$$\begin{cases} V_1(t-\tau) + \dfrac{Z_c}{2} I_0(t) = \dfrac{V_0(t)}{2} + \dfrac{V_0(t-2\tau)}{2} + \dfrac{Z_c}{2} I_0(t-2\tau) \\ I_1(t-\tau) + \dfrac{I_0(t)}{2} = -\dfrac{V_0(t)}{2Z_c} + \dfrac{V_0(t-2\tau)}{2Z_c} + \dfrac{I_0(t-2\tau)}{2} \end{cases}$$

2nd segment
$$\begin{cases} V_2(t-2\tau) - V_1(t-\tau) + \dfrac{Z_c}{2} I_{12}(t-\tau) = \dfrac{V_1(t-3\tau)}{2} + \dfrac{Z_c}{2} I_{12}(t-3\tau) \\ I_2(t-2\tau) + \dfrac{V_1(t-\tau)}{4Z_c} - \dfrac{I_{12}(t-\tau)}{2} = -\dfrac{V_1(t-3\tau)}{4Z_c} + \dfrac{I_{12}(t-3\tau)}{2} \\ V_2'(t-2\tau) = V_2(t-2\tau) \\ V_2(t-2\tau) = Z_c I_2(t-2\tau) \end{cases}$$

3rd segment
$$\begin{cases} V_3(t-2\tau) - V_1(t-\tau) + \dfrac{Z_c}{2} I_{13}(t-\tau) = \dfrac{V_1(t-3\tau)}{2} + \dfrac{Z_c}{2} I_{13}(t-3\tau) \\ I_3(t-2\tau) + \dfrac{V_1(t-\tau)}{4Z_c} - \dfrac{I_{13}(t-\tau)}{2} = \dfrac{V_1(t-3\tau)}{4Z_c} + \dfrac{I_{13}(t-3\tau)}{2} \\ V_3'(t-2\tau) = V_3(t-2\tau) \\ V_3(t-2\tau) = Z_c I_3(t-2\tau) \end{cases}$$

According to Kirchoff's law, $I_1(t-\tau) = I_{12}(t-\tau) + I_{13}(t-\tau)$.

Consider the following time steps.

Any function like V_i and I_i has to satisfy

$$\begin{cases} f(t-\tau) = 0 & \text{for } t \le \tau \\ f(t-\tau) = 0 & \text{for } t > \tau \end{cases}$$

For $t < \tau$ we obtain

$$I_0(t) = \frac{V_0(t)}{Z_c}$$

and

$$V_1(t-\tau) = I_i(t-\tau) = 0$$

The input voltage does not reach the first segment terminal where the bifurcation occurs.

For $\tau < t < 2\tau$ we obtain

$$I_0(t) = \frac{V_0(t)}{Z_c}$$
$$V_1(t) = V_0(t-\tau)$$
$$I_1(t) = \frac{V_0(t-\tau)}{Z_c}$$
$$V_2(t) = V_3(t) = 0$$

The bifurcation has been traveled by the input voltage, which does not meet the other segment terminals yet, thus creating a first reflected wave towards the first segment input with a reflection coefficient equal to 1/3 because the load seen at the bifurcation node equivalent to the similar segments 2 and 3 is equal to $Z_c/2$.

For $2\tau < t < 3\tau$ all the line terminals are reached by the input voltage:

$$I_0(t) = \frac{V_0(t)}{Z_c}$$
$$V_1(t) = V_0(t-\tau)$$
$$I_1(t) = \frac{V_0(t-\tau)}{Z_c}$$
$$V_2(t) = V_3(t) = V_0(t-2\tau)$$

For $3\tau < t < 5\tau$ while a new reflection occurs at the bifurcation node

$$V_1(t) = V_0(t-\tau) - \frac{V_0(t-3\tau)}{3}$$

towards the input, we obtain

$$I_1(t) = \frac{V_0(t-\tau)}{Z_c} + \frac{V_0(t-3\tau)}{3Z_c}$$

Bifurcation : 2Zc2=2Zc3=Zc1=200 Zin2=Zin3=100 l=7.5cm

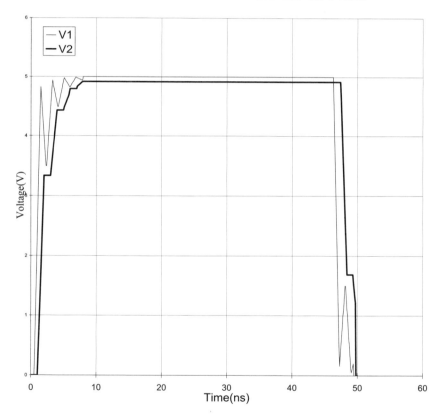

Figure 5.14. *Waveforms in front of, then behind, the bifurcation drawn in Figure 5.13*

The term $V_0(t-3\tau)/3$ comes from the total reflection on the first segment input of the first reflected wave on the bifurcation node.

Segments 2 and 3 being strictly similar, they have the same voltage equal to

156 Electromagnetism and Interconnections

$$V_2(t) = \frac{2}{3}V_0(t-2\tau)$$

and so on.

5.2.4. *Complex distributed parameter networks*

In applications, it is necessary to find a suitable numbering of all the coupled segments in respect of the chronology of pulses propagation along them in spite of the fact that they have different lengths and several coupling areas as well. This numbering, based on the topological analysis of electromagnetic interferences developed in Chapter 2, is needed for chaining all the matrix equations attached to the coupled transmission lines, to the purely resistive lumped networks, and to the bifurcations.

After this, we can create the *state vectors* [*X*] as

$$\begin{bmatrix} \vdots \\ [X^K] \\ \vdots \end{bmatrix}$$

where the sub-vector [X^K] corresponds to a set of coupled line segments which is itself divided in three parts:

$$\text{sub-vector tied to number } k \begin{cases} \begin{bmatrix} [\,] \end{bmatrix} \} \text{ part related to input nodes of lines } k \\ \begin{bmatrix} [\,] \end{bmatrix} \} \text{ part related to total coupled lines } k \\ \begin{bmatrix} [\,] \end{bmatrix} \} \text{ part related to output nodes of lines } k \end{cases}$$

Each part is divided itself into voltage lines followed by current lines. Two adjacent sub-vectors are tied together by a linear *state equation* corresponding to the transmission line, lumped elements, and bifurcations equations using the matrix block form.

Such *state equations* can be written as

$$[A]\left[X\left(t-\frac{[z]}{v}\right)\right] = \left[Y\left(t-\frac{[z_0]}{v}\right)\right] + [B]\left[X\left(t-\frac{2[z]}{v}\right)\right] \qquad [5.26]$$

where [*A*] and [*B*] are square matrices having structures like those of the examples shown in Figures 5.15 and 5.16.

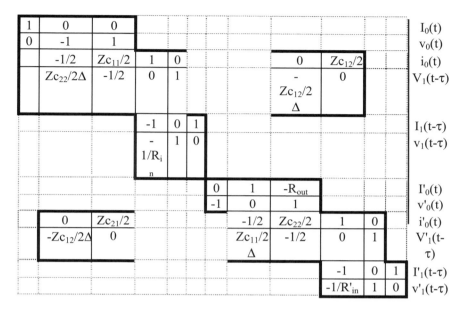

Figure 5.15. *Structure of matrix [A] in the case of the crosstalk of Figure 5.6*

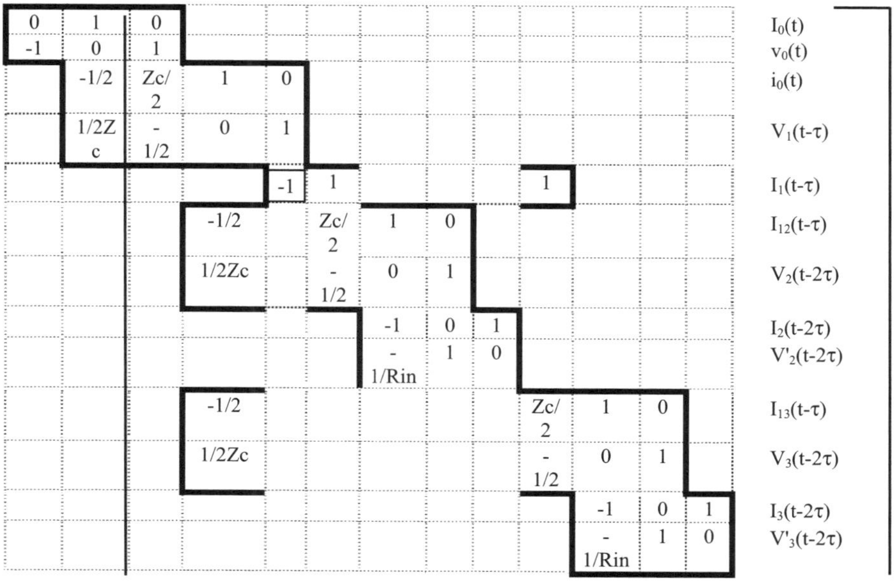

Figure 5.16. *Structure of the matrix [A] in the case of the bifurcation of Figure 5.13*

– $[z]$ is the row vector containing the distances traveled along each conductor from its origin up to the point of interest, each line of which corresponds to the same line number of the vector $[X]$. Velocity v is the common velocity of the pulses on all the lines.

– $\left[X\left(t - \dfrac{[z]}{v} \right) \right]$ is the vector of the unknown voltage and current vector at the latest past time with suitable shifts to account for the propagation synchronization already mentioned.

– $\left[X\left(t - \dfrac{2[z]}{v} \right) \right]$ is the voltage and current vectors already known because they come from reflections.

– $\left[Y\left(t - \dfrac{[z_0]}{v} \right) \right]$ is the known vector of voltage or current sources at a point defined by the row vector $[z_0]$ of which each line is tied to the corresponding line of $[Y]$.

Then, equation [5.26] is solved in respect of the unknown vector

$$\left[X\left(t - \dfrac{[z]}{v} \right) \right]$$

by means of the Gauss algorithm which has to be used only once in the case of purely resistive linear lumped elements and as many times as necessary in the case of nonlinear lumped elements not depending on the frequency.

How do we estimate the computation charge related to the above process?

In order to answer this, let L be the number of transmission line segments and choose the Gauss algorithm. This requires $(1/3)(4L)^3$ basic operations like additions and multiplications.

Furthermore, it is necessary to store $2[(4L)^2 + 4L]$ numbers in the system memory.

Appendix D shows a modified Gauss-Seidel algorithm which is a recursive method of solving [5.26] by saving a lot of storage capacity.

5.2.5. *Estimation of the transient state time of signals*

In order to estimate the transient state time of signals leaving a complex interconnection network fed by source signals being zero before $t = 0$ and equal to 1 afterwards, we have to go further with the recursive equation, thus finding a criterion of convergence. Then, let us write the previous state equation as follows:

$$\left[X\left(t-\frac{[z]}{v}\right)\right] = \underbrace{[A]^{-1}\left[Y\left(t-\frac{[z_0]}{v}\right)\right]}_{[Z(t)]} + \underbrace{[A]^{-1}[B]}_{[C]}\left[X\left(t-\frac{2[z]}{v}\right)\right]$$

Thus, after n times, this yields

$$\left[X\left(t-\frac{[z]}{v}\right)\right] = \sum_{k=0}^{k=n-1}[C]^k[Z(t)] + [C]^n\left[X\left(t-\frac{2[z]}{v}\right)\right] \quad \text{towards} \quad \sum_{k=0}^{\infty}[C]^k[Z(t)]$$

Now we deal with the norms of matrices. Accordingly, firstly the triangle inequality $\|A+B\| \leq \|A\| + \|B\|$ then, secondly, the inequality $\|AB\| \leq \|A\|\|B\|$ easily leads to

$$\left\|\left[X\left(t-\frac{[z]}{v}\right)\right] - \sum_{k=0}^{k=n-1}[C]^k[Z(t)]\right\| = \left\|\sum_{k=n}^{\infty}[C]^k[Z(t)]\right\| \leq \frac{\|C\|^n \|Z(t)\|}{1-\|C\|}$$

which shows that the modulus of the matrix $[C] = \{[A]\}^{-1}\{[B]\}$ has to be less than 1. From any maximum value of the modulus $[C]$, we immediately obtain the value of n corresponding to a given deviation:

$$\frac{\left\|\left[X\left(t-\frac{[z]}{v}\right)\right] - \sum_{k=0}^{k=n-1}[C]^k[Z(t)]\right\|}{\|Z(t)\|} \leq \frac{\|C\|^n}{1-\|C\|} \leq \varepsilon$$

which defines the *transient state time* $n\tau$ of the outgoing signals in respect of the incoming signals which are zero before $t = 0$ and equal to 1 afterwards.

It is possible to go further by means of the theory of the spectral expansion of matrices [LAS3]. According to this theory, any square matrix having the dimensions $n \times n$, like $[C]$, and different eigenvalues λ_m, can be expanded as follows:

$$[C] = \sum_{m=1}^{m=n} \lambda_m [P_m]$$

where $[P_m]$ are particular matrices called "projectors".

These properties are

$$\begin{cases} [P_m]^2 = [P_m] \\ [P_m][P_1] = [0] \quad \text{if} \quad m \neq 1 \end{cases}$$

According to the general Duncan-Sylvester formula, these projectors are expressed as

$$P_m = \prod_{1 \neq m} \frac{[A] - \lambda_1 [1]}{\lambda_m - \lambda_1}$$

Let us return to the state matrix equation written as

$$\left[X\left(t - \frac{[z]}{v} \right) \right] = [Z(t)] + [C] \left[X\left(t - \frac{2[z]}{v} \right) \right]$$

We multiply it on its left-hand side by the projector $[P_j]$. Accounting for the properties of projectors and the formula of the spectral expansion of the matrix $[C]$, we obtain

$$\underbrace{[P_j]\left[X\left(t - \frac{[z]}{v} \right) \right]}_{\left[X_j\left(t - \frac{[z]}{v} \right) \right]} = \underbrace{[P_j][Z(t)]}_{[Z_j(t)]} + \underbrace{[P_j][C]}_{\lambda_j [P_j]} \underbrace{\left[X\left(t - \frac{2[z]}{v} \right) \right]}_{\lambda_j \left[X_j\left(t - \frac{2[z]}{v} \right) \right]}$$

Therefore, this yields a lot of single transmission line equations devoted to the eigenvalues λ_j of the matrix $[C]$: this is the *diagonalization process* of the coupled transmission line matrix equation. The interest of this process lies in highlighting the dynamic behavior of the signals.

Indeed, when an eigenvalue is positive, the corresponding signal is monotonically increasing, while if it is negative, then ringing around the asymptotic value occurs.

Direct Time-domain Methods   161

We should consider carefully the highest modulus of the eigenvalues which is called the *spectral radius* $\rho([C])$ of the matrix $[C]$. This spectral radius is the upper bound of any norm of the matrix $[C]$. Then, the transient state time $n\tau$ is related to this spectral radius by means of

$$\frac{\{\rho[C]\}^n}{1-\rho[C]} \leq \varepsilon$$

5.2.6. *Numerical computation of the characteristic impedance matrix*

In section 5.2.1, concerning homogenous coupling, the characteristic impedance matrix has been defined as

$$[Z_{c0}] = \frac{1}{v}[C]^{-1}$$

According to the topological analysis developed in section 2.3, it is possible to classify all the electromagnetic interferences so that $[C]$ becomes almost diagonal. Physically and accounting for the solutions of Laplace's equation, the main diagonal terms of this capacitor matrix are k times greater than those of the two adjacent parallel diagonals. These are k times greater than the other ones lying on the parallel exterior diagonals, and so on.

Practically, matrix $[C]$ can be considered as tridiagonal with an error of less than 10^{-3}.

Let c_{ii} be the main diagonal terms of $[C]$. It is always possible to write

$$[C] = [c_{ii}]\{[1]-[\varGamma]\}$$

where all the terms of $[\varGamma]$ are non-negative.

Thus, we obtain the matrix expansion

$$[C]^{-1} = \{[1]+[\varGamma]+[\varGamma]^2 + \cdots\}[c_{ii}]^{-1}$$

which converges very quickly as $(1/k)^n$.

5.3. Conclusion

In the case of transmission lines having coupled lossless conductors, the matrix equations found in the previous chapter are solved directly by means of the spectral theory of matrices. We then show the existence of a linear "propagator" which leads naturally to a recursive equation for multiple reflection modeling. This equation has been extended to a general matrix state equation for formulating the electrical behavior of any complex distributed parameter network.

The computation time and storage memory required to obtain the amplitude of many times reflected waveforms at any time are the lowest compared to any other method, but with the constraint of accounting for neither losses nor lumped elements between the segments of lines.

Chapter 6

Discretization in the Time Domain

The second method of solving the equations of propagation along interconnections (the first being the subject of Chapter 5) is well known according to the name *finite differences* and is necessary for modeling the electrical behavior of lossy or lossless lumped elements tied to electronic devices at line terminals. Then, the further terms related to the *finite differences* in the time domain are linked to the previous *matrix state equation*.

The third method, called the *velocity operator interpolation* method, is developed for the full wave analysis of straight heterogenous stratified media and is completely independent of the other two methods.

6.1. Finite difference method in the time domain

6.1.1. *From full wave analysis to nodal operational matrices*

It has been mentioned previously that full wave analysis is not well suited for modeling a complete interconnection network, a few elements of which are depicted in Figure 2.2 of section 2.1.3. However, it has been noted that this could be applied at some particular points, which we mention here.

In Figure 6.1, we can see at each terminal of the transmission lines elements such as:

164   Electromagnetism and Interconnections

– geometrical discontinuities within transmission lines like "vias", "turns", "width rough variations", etc., and

– external connections of passive or active components with their packaging interconnections.

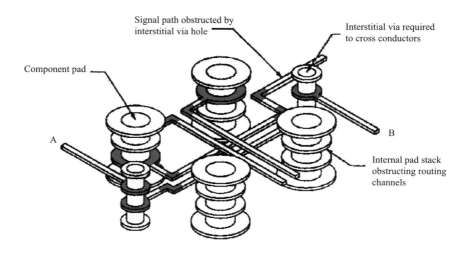

Figure 6.1. *Elements generating nodal matrices (gray shaded areas)*

These cannot be modeled by means of transmission line equations because they obviously have a 3D structure.

Let us come back to the moment methods which are involved in full wave analysis (section 2.1.3). The electromagnetic fields $\vec{E}$ and $\vec{H}$ are expanded in any base made of known functions $\vec{E}_i(x,y,z)$, $\vec{H}_i(x,y,z)$ with their coefficients depending on time $e_i(t)$, $h_i(t)$. Assuming zero initial conditions, we obtain the matrix equation [2.6]:

$$[M]\frac{\partial}{\partial t}\begin{bmatrix}[e]\\[h]\end{bmatrix}+[N]\begin{bmatrix}[e]\\[h]\end{bmatrix}+[R]=[0]$$

where the transposed matrix of $[e]$ is $[e_1(t), e_2(t),\ldots, e_I(t)]$ and that of $[h]$ is $[h_1(t), h_2(t),\ldots, h_I(t)]$:

$$[M] = \begin{bmatrix} -[(\varepsilon \vec{A}_j, \vec{E}_i)] & [0] \\ [0] & [(\mu \vec{A}_{mj}, \breve{\vec{H}}_i)] \end{bmatrix}$$

$$[N] = \begin{bmatrix} -[(\sigma \vec{A}_j, \vec{E}_i)] & [(\vec{\nabla} \wedge \vec{A}_j, \breve{\vec{H}}_i)] \\ [(\vec{\nabla} \wedge \vec{A}_{mj}, \vec{E}_i)] & [0] \end{bmatrix}$$

$$[R] = \underbrace{\begin{bmatrix} \left[\iint_{\partial\Omega} \vec{j}_s \vec{A} \, \mathrm{d}s \right] \\ \left[\iint_{\partial\Omega} \vec{j}_{ms} \vec{A}_m \, \mathrm{d}s \right] \end{bmatrix}}_{\text{boundary conditions}}$$

Let us share the domain Ω of the electromagnetic propagation around a given transmission line terminal into small cubes having their edges parallel to the Cartesian axes. The vector test functions $\vec{A}$ and $\vec{A}_m$ of the method of moments are chosen like unitary vectors $\vec{u}$ and $\vec{u}_m$, respectively, collinear to each edge of the cube. The volume and the external area of cube number i are denoted ω_i and s_i, respectively, and its edge length is equal to a_i.

So, the scalar products $(\varepsilon \vec{A}_j, \vec{E}_i)$, $(\sigma \vec{A}_j, \vec{E}_i)$ and $(\mu \vec{A}_{mj}, \breve{\vec{H}}_i)$ become $\varepsilon \vec{u} \vec{E}_i \omega_i$, $\sigma \vec{u} \vec{E}_i \omega_i$ and $\mu \vec{u}_m \vec{H}_i \omega_i$, respectively, and are considered to belong to the input matrix of the model.

Let $\sigma_{\vec{A}_j} = \vec{u}$ and $\sigma_{\vec{A}_{mj}} = \vec{u}_m$ be the jumps of the test functions $\vec{A}$ and $\vec{A}_m$, respectively, across the faces of any cube. Note that $\vec{\nabla} \wedge \vec{A}_j$ and $\vec{\nabla} \wedge \vec{A}_{mj}$ become the singular distributions on the external faces of each finite element $\vec{n} \wedge \sigma_{\vec{A}_j} \delta_S$ and $\vec{n} \wedge \sigma_{\vec{A}_{mj}} \delta_S$, respectively. So, inside matrix $[N]$, the terms $(\vec{\nabla} \wedge \vec{A}_j, \breve{\vec{H}}_i) = \vec{u}(\vec{H}_i \wedge \vec{n})s_i$ and $(\vec{\nabla} \wedge \vec{A}_{mj}, \vec{E}_i) = \vec{u}_m(\vec{E}_i \wedge \vec{n})s_i$ involve electrical and magnetic current densities $-(\vec{H}_i \wedge \vec{n})$ and $-(\vec{E}_i \wedge \vec{n})$, respectively: they have to leave this matrix to go into the other matrix $[R]$ which becomes the output matrix of the model. Let

$$\vec{E}_i^{\bullet}(x,y,z,t) = e_i(t)\vec{E}_i(x,y,z)$$
$$\vec{\breve{H}}_i^{\bullet}(x,y,z,t) = h_i(t)\vec{\breve{H}}_i(x,y,z)$$

inside each finite element being a small cube. Accounting for the matrix [R], this yields the equations

$$\left(\sigma + \varepsilon \frac{\partial}{\partial t}\right)\left[\vec{u}\vec{E}_i^{\bullet}(x,y,z,t)\omega_i\right] = \vec{u}_i(\vec{H}_i \wedge \vec{n} + \vec{j})s_i$$

$$\mu \frac{\partial}{\partial t}\left[\vec{u}_m \vec{\breve{H}}_i^{\bullet}(x,y,z,t)\omega_i\right] = \vec{u}_m(\vec{E}_i \wedge \vec{n} + \vec{j}_{ms})s_i$$

In order to research a classical lumped electrical model, accounting for the very small size of the cubes, it remains to consider firstly the circulation of the electrical field along an edge of cube i:

$$\int_{edge} \vec{E} \, d\vec{l} = a_i \times (\text{projection of } \vec{E} \text{ onto the edge})$$

gives the scalar potential difference $V_{i,k} - V_{i,k+1}$ between the two nodes k and $k+1$ of this edge. Secondly, we consider the circulation of the magnetic field around a face of the finite element:

$$\int_{face} \vec{\breve{H}} \, d\vec{l}' = a_i \times (\text{sum of projections of } \vec{\breve{H}} \text{ onto the edges})$$

gives the difference between mesh electrical currents $J_{i,l} - J_{i+1,l}$ crossing face l adjacent to cubes i and $i+1$.

Then, replacing the coordinates of the electromagnetic fields involved in the last matrix equation by their corresponding voltages and currents, leads to a new matrix equation as regards only the electrical currents and potentials:

$$\begin{bmatrix}[V]\\{}[J]\end{bmatrix}_{out} = \left\{\underbrace{\begin{bmatrix}[0] & [L]\\{}[C] & [0]\end{bmatrix}\frac{\partial}{\partial t} + \begin{bmatrix}[0] & [R]\\{}[0] & [0]\end{bmatrix}}_{\text{nodal operational matrix}}\right\}\begin{bmatrix}[V]\\{}[J]\end{bmatrix}_{in} \qquad [6.1]$$

This defines the first-order nodal operational matrix tied to the 3D modeling of an interconnection element.

This is the matrix equation of a lumped electrical network defined as made of self- and mutual inductances and capacitances which are electrical parameters that do not depend on any space coordinate.

In this case, the modeling of lumped elements is based on the use of quadripoles in series, as shown in Figure 6.2, where their second-order nodal operational matrices are shown.

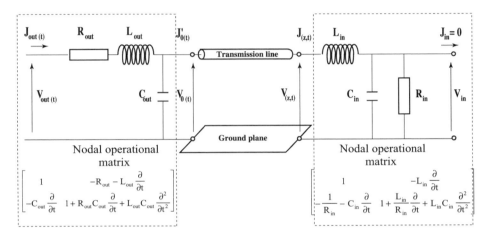

Figure 6.2. *Input and output quadripoles and nodal operational matrices*

6.1.2. Recursive differential transmission line matrix equation of complex networks

In section 5.2.4, we formulated the *state vectors* $[X]$ as

$$\begin{bmatrix} \vdots \\ [X^K] \\ \vdots \end{bmatrix}$$

where the sub-vector $[X^K]$ is divided into three parts:

sub-vector number k $\begin{cases} [\,] \} \text{ lumped elements: differential matrix equation} \\ [\,] \} \text{ transmission line: linear recursive matrix equation} \\ [\,] \} \text{ lumped elements: differential matrix equation} \end{cases}$

Each part is divided itself into voltage lines followed by current lines.

168   Electromagnetism and Interconnections

Then, introducing the lumped models of nodes between two adjacent transmission lines changes matrix state equation [5.26] introduced at the end of Chapter 5

$$[A]\left[X\left(t-\frac{[z]}{v}\right)\right] = \left[Y\left(t-\frac{[z_0]}{v}\right)\right] + [B]\left[X\left(t-\frac{2[z]}{v}\right)\right]$$

into a new one being both recursive and differential in the time domain:

$$\left\{[A]+[A']\frac{\partial}{\partial t}\right\}\left[X\left(t-\frac{[z]}{v}\right)\right] = \left[Y\left(t-\frac{[z_0]}{v}\right)\right] + \left\{[B]+[B']\frac{\partial}{\partial t}\right\}\left[X\left(t-\frac{2[z]}{v}\right)\right]$$

[6.2]

which we call the *recursive-differential equation* which has to be solved in practical applications of the industrial field.

6.1.3. *Estimation of the transient state time*

Note that the operator $[A]+[A'](\partial/\partial t)$ is invertible. The differential matrix equation

$$\left\{[A]+[A']\frac{\partial}{\partial t}\right\}[X(t)] = [Y(t)]$$

has a solution easily obtained by means of the transformation

$$[X(t)] = \exp\{[A']^{-1}[A]t\}[U(t)]$$

So, we can write the recursive-differential equation as highlighting two new matrices $[Z]$ and $[C]$:

$$\left[X\left(t-\frac{[z]}{v}\right)\right] = \underbrace{\left\{[A]+[A']\frac{\partial}{\partial t}\right\}^{-1}\left[Y\left(t-\frac{[z_0]}{v}\right)\right]}_{[Z(t)]} +$$

$$\underbrace{\left\{[A]+[A']\frac{\partial}{\partial t}\right\}^{-1}\left\{[B]+[B']\frac{\partial}{\partial t}\right\}}_{\left[C\left(\frac{\partial}{\partial t}\right)\right]}\left[X\left(t-\frac{2[z]}{v}\right)\right]$$

In the same way as in Chapter 5, we straightforwardly go to the recurrence in order to find the criterion of convergence:

$$\left[X\left(t-\frac{[z]}{v}\right)\right] = \sum_{k=0}^{k=n-1}\left[C\left(\frac{\partial}{\partial t}\right)\right]^k [Z(t)] + \left[C\left(\frac{\partial}{\partial t}\right)\right]^n \left[X\left(t-\frac{2[z]}{v}\right)\right]$$

towards $\sum_{k=0}^{\infty}\left[C\left(\frac{\partial}{\partial t}\right)\right]^k [Z(t)]$

when n becomes infinite.

Here, we have to use the norms of both matrices and operators in the framework of the theory of linear operators belonging to the set called a Banach space at the same time. The definition of the norm of a linear operator A acting in a functional space E made up of functions f is

$$\|A\| = \sup_{f \in E} \frac{\|A \circ f\|}{\|f\|}$$

According to firstly the triangle inequality $\|A+B\| \leq \|A\| + \|B\|$ and secondly the inequality $\|AB\| \leq \|A\|\|B\|$, the following inequalities occur:

$$\left\|\left[X\left(t-\frac{[z]}{v}\right)\right] - \sum_{k=0}^{k=n-1}\left[C\left(\frac{\partial}{\partial t}\right)\right]^k [Z(t)]\right\| = \left\|\sum_{k=n}^{\infty}\left[C\left(\frac{\partial}{\partial t}\right)\right]^k [Z(t)]\right\| \leq \frac{\left\|C\left(\frac{\partial}{\partial t}\right)\right\|^n \|[Z(t)]\|}{1 - \left\|C\left(\frac{\partial}{\partial t}\right)\right\|}$$

which shows that the modulus of $[C]$ has to be less than 1. Indeed, its definition leads to

$$\left[C\left(\frac{\partial}{\partial t}\right)\right] = \left\{[A] + [A']\frac{\partial}{\partial t}\right\}^{-1}\left\{[B] + [B']\frac{\partial}{\partial t}\right\}$$

Applying the inequality $\|A+B\| \geq \|\|A\| - \|B\|\|$ in the first term and the triangle inequality $\|A+B\| \leq \|A\| + \|B\|$ in the second term, we obtain an upper value of the norm of $[C]$ to which n is related in respect of a given deviation:

170   Electromagnetism and Interconnections

$$\frac{\left\|\left[X\left(t-\frac{[z]}{v}\right)\right]-\sum_{k=0}^{k=n-1}\left[C\left(\frac{\partial}{\partial t}\right)\right]^{k}[Z(t)]\right\|}{\|[Z(t)]\|} \leq \frac{\left\|\left[C\left(\frac{\partial}{\partial t}\right)\right]\right\|^{n}}{1-\left\|\left[C\left(\frac{\partial}{\partial t}\right)\right]\right\|} \leq \varepsilon$$

which defines the transient state time $n\tau$ of the outgoing signals corresponding to an incoming Heaviside step that is an infinitely straight edge waveform.

6.1.4. Finite difference approximation of differential operators in the time domain

Now we deal with the basics of the *finite difference method*. Let us recall the well-known Taylor formula:

$$f(t+h) = f(t) + hf'(t) + \frac{h^2}{2} f''(t) + \cdots + \frac{h^n}{n!} f^{(n)}(t) + \cdots$$

Let the time shift operator be defined as follows:

$$f(t) \xrightarrow{\tau_h} f(t-h) = \tau_h[f(t)] = (\tau_h \circ f)(t)$$

leading us to write both parts of the Taylor formula in an operational form:

$$\tau_{-h} \circ f = 1 \circ f + h \frac{d}{dt} \circ f + \frac{h^2}{2} \frac{d^2}{dt^2} \circ f + \cdots + \frac{h^n}{n!} \frac{d^n}{dt^n} \circ f$$

In accordance with the convergence of the right-hand side of the expansion, this involves

$$\left| \frac{h^n}{n!} \frac{d^n}{dt^n} \circ f \right| \to 0, n \to \infty$$

that is met when $f \in C^\infty$, a set of infinitely differentiable functions. In the case of convergence, we can write the operational relation:

$$\tau_{-h} = \exp\left(h \frac{d}{dt}\right)$$

Reciprocally, we obtain

$$\frac{d}{dt} = \frac{1}{h}\ln\left[(\tau_{-h}-1)+1\right] = \frac{1}{h}\sum_{n=1}^{\infty}(-1)^{n-1}\frac{(\tau_{-h}-1)^n}{n}$$

which defines *Newton's down scheme*, in so far as the so-called projection operator made up of the first N elements of the Taylor expansion

$$A_{h,N} = \frac{1}{h}\sum_{n=1}^{N}(-1)^{n-1}\frac{(\tau_{-h}-1)^n}{n}$$

converges towards d/dt when $n \to \infty$ in the framework of the theory of linear operators belonging to the set called a Banach space.

In order to get the required condition of convergence, we do work on the norm of the operators. According to the definition of the norm of a linear operator A acting in a functional space E made up of functions f:

$$\|A\| = \sup_{f \in E} \frac{\|A \circ f\|}{\|f\|}$$

we obtain the following criterion of convergence:

$$\left\|\frac{d}{dt} - A_{h,N}\right\| = \frac{1}{h}\left\|\sum_{n=N+1}^{\infty}(-1)^{n-1}\frac{(\tau_{-h}-1)^n}{n}\right\| \leq \frac{1}{h}\sum_{n=N+1}^{\infty}\frac{\|(\tau_{-h}-1)^n\|}{n}$$

according to firstly the triangle inequality $\|A+B\| \leq \|A\| + \|B\|$ and secondly the inequality $\|AB\| \leq \|A\|\|B\|$.

Let $n = N + m$, then $1/n < 1/m$ and we can write

$$\left\|\frac{d}{dt} - A_{h,N}\right\| \leq \frac{1}{h}\|(\tau_{-h}-1)\|^N \sum_{m=1}^{\infty}\frac{1}{m}\|(\tau_{-h}-1)\|^m = \frac{1}{h}\|(\tau_{-h}-1)\|^N \ln\left|-1+\|(\tau_{-h}-1)\|\right|$$

The convergence is met since $\|(\tau_{-h}-1)\| \leq Kh^\nu$, which defines the so-called *Hölderian property* of the signal which has to be merely continuous in the case $\nu = 1$. Therefore, we have to satisfy

$$\frac{1}{h}(Kh^v)^N \ln(1-Kh^v) < \varepsilon$$

In the same manner, *Newton's up scheme* can be defined by means of the choice

$$\tau_h = \exp\left(-h\frac{d}{dt}\right)$$

then

$$\frac{d}{dt} = -\frac{1}{h}\ln[1-(1-\tau_h)] = \frac{1}{h}\sum_{n=1}^{\infty}\frac{(1-\tau_h)^n}{n}$$

with the following criterion of convergence:

$$\left\|\frac{d}{dt}-A_{h,N}\right\| \leq \frac{1}{h}\|(1-\tau_h)\|^N \sum_{m=1}^{\infty}\frac{1}{m}\|(1-\tau_h)\|^m = \frac{1}{h}\|(1-\tau_h)\|^N \ln\left|-1+\|(1-\tau_h)\|\right|$$

The convergence is reached since $\|1-\tau_h\| \leq Kh^v$, with $v > 1/(N+1)$, which means the condition to be satisfied is

$$\frac{1}{h}(Kh^v)^N \ln(1-Kh^v) < \varepsilon$$

The last scheme is called *Newton's centered scheme* with the choice

$$\frac{\tau_{-h}-\tau_h}{2} = \frac{1}{2}\left[\exp\left(h\frac{d}{dt}\right)-\exp\left(-h\frac{d}{dt}\right)\right] = sh\left(h\frac{d}{dt}\right)$$

from which arises the Taylor expansion versus $\frac{\tau_{-h}-\tau_h}{2}$:

$$\frac{d}{dt} = \frac{1}{h}\operatorname{Arg}sh\left(\frac{\tau_{-h}-\tau_h}{2}\right) = \frac{1}{h}\left[\frac{\tau_{-h}-\tau_h}{2}+\frac{(-1)^n}{(2n+1)}\frac{1,3,5,\ldots(2n-1)}{2,4,6,\ldots 2n}\left(\frac{\tau_{-h}-\tau_h}{2}\right)^{2n+1}\right]$$

with the criterion of convergence based on finding the maximum value of

$$\left\|\frac{d}{dt} - A_{h,N}\right\| = \frac{1}{h}\left\|\sum_{n=N+1}^{\infty}\frac{(-1)^n}{(2n+1)}\frac{1,3,5,\ldots(2n-1)}{2,4,6,\ldots 2n}\left(\frac{\tau_{-h}-\tau_h}{2}\right)^{2n+1}\right\|$$

$$\leq \frac{1}{h}\sum_{n=N+1}^{\infty}\frac{1}{2n+1}\frac{1,3,5,\ldots(2n-1)}{2,4,6,\ldots 2n}x^{2n+1}$$

where

$$x = \left\|\frac{\tau_{-h}-\tau_h}{2}\right\|$$

Let $n = N + m$, m varying from 1 to ∞, so sharing the internal fraction into two parts yields

$$\left\|\frac{d}{dt} - A_{h,N}\right\| < x^{2N}\left\{\sum_{m=1}^{\infty}\frac{1,3,5,\ldots(2m-1)}{2,4,6,\ldots 2m}\underbrace{\left[\frac{(2m+1)\ldots(2n-1)}{(2m+2)\ldots(2n)}\right]}_{<1}\frac{x^{2m+1}}{(2m+1)}\right\}$$

$$< x^{2N}\left\{\sum_{m=1}^{\infty}\frac{1,3,5,\ldots(2m-1)}{2,4,6,\ldots 2m}\frac{x^{2m+1}}{(2m+1)}\right\} = \frac{1}{h}\left\|\frac{\tau_{-h}-\tau_h}{2}\right\|^{2N}\left[\sin^{-1}\left\|\frac{\tau_{-h}-\tau_h}{2}\right\| - \left\|\frac{\tau_{-h}-\tau_h}{2}\right\|\right]$$

The convergence is reached since $\|\tau_{-h}-\tau_h\| \leq Kh^v$, with $v > 1/(2N + 3)$, which means the condition to be satisfied is

$$\frac{1}{h}(Kh^v)^N\left[\sin^{-1}(Kh^v) - (Kh^v)\right] < \varepsilon$$

6.1.5. *Application to lumped quadripole modeling approximation in the time domain*

Let us return to the electrical scheme of Figure 6.2. The approximation of the linear operators involved inside the nodal matrices according to the finite difference method in the time domain is based on the *projection operator* $A_{\Delta t,1}$:

$$\frac{\partial}{\partial t} \xrightarrow{A_{\Delta t,1}} A_{\Delta t,1}\left(\frac{\partial}{\partial t}\right) = (\tau_0 - \tau_{\Delta t})\frac{1}{\Delta t} \qquad [6.3]$$

In practice, Δt is chosen as less than the smallest response time met in the elements of the operational matrices. It is called the *time step* of the discretization process.

Let us consider the input quadripole when $L_{out} = 0$ with its input set in short circuit and its output in open circuit. Creating a current pulse like the Dirac distribution at the input, we obtain the current $e^{-t/R_{out}C_{out}}$ flowing out of the resistor. Now approximating the differential equation:

$$\left(1 + R_{out}C_{out}\frac{\partial}{\partial t}\right)(t) = 0$$

leading to this current by means of the finite difference method, we obtain a recursive relation giving the time discrete solution:

$$\left(1 + \frac{\Delta t}{R_{out}C_{out}}\right)^{-n} = \exp\left[-n\log\left(1 + \frac{\Delta t}{R_{out}C_{out}}\right)\right]$$

$$= \exp\left[-n\frac{\Delta t}{R_{out}C_{out}}\left(1 + \frac{\Delta t}{2R_{out}C_{out}} + (\Delta t)^2\right)\right]$$

which gives the constraint $\Delta t < R_{out}C_{out}$.

In the case of the MOS technology of semiconductors, this time is the shortest one.

The computation process proceeds as follows. We recall the multiple reflections recursive equation:

$$\begin{bmatrix} \frac{1}{2} & \frac{Z_{c0}}{2} \\ \frac{1}{2Z_{c0}} & \frac{1}{2} \end{bmatrix} \begin{bmatrix} V_0(z,t) \\ J_0(z,t) \end{bmatrix} = \begin{bmatrix} V\left(t - \frac{z}{v}\right) \\ J\left(t - \frac{z}{v}\right) \end{bmatrix} - \begin{bmatrix} \frac{1}{2} & \frac{-Z_{c0}}{2} \\ \frac{-1}{2Z_{c0}} & \frac{1}{2} \end{bmatrix} \begin{bmatrix} V_0\left(t - \frac{2z}{v}\right) \\ J_0\left(t - \frac{2z}{v}\right) \end{bmatrix} \quad [6.4]$$

which has to be completed by the nodal equations at both the input and output terminals where first the transformation $A_{\Delta t}$ is applied, then operations of time shifts have to be carried out in order to account for the synchronization of the events. This gives the following result:

$$\underbrace{\begin{bmatrix} V_0\left(t-\dfrac{z}{v}\right) \\ J_0\left(t-\dfrac{z}{v}\right) \end{bmatrix}}_{\text{transmission line input}} = \sum_{k\geq 0} \underbrace{[T_{in,k}]}_{\substack{\text{external} \\ \text{connections}}} \underbrace{\begin{bmatrix} V_{out}\left(t-\dfrac{z}{v}-k\Delta t\right) \\ J_{out}\left(t-\dfrac{z}{v}-k\Delta t\right) \end{bmatrix}}_{\text{active device output}} \qquad [6.5]$$

where the matrices corresponding to the external connections are expressed as follows:

$$[T_{in,0}] = \begin{bmatrix} 1 & -R_{out} - \dfrac{L_{out}}{\Delta t} \\ -\dfrac{C_{out}}{\Delta t} & 1 + \dfrac{R_{out}C_{out}}{\Delta t} + \dfrac{L_{out}C_{out}}{\Delta t^2} \end{bmatrix}$$

$$[T_{in,1}] = \begin{bmatrix} 1 & \dfrac{L_{out}}{\Delta t} \\ \dfrac{C_{out}}{\Delta t} & -\dfrac{R_{out}C_{out}}{\Delta t} - 2\dfrac{L_{out}C_{out}}{\Delta t^2} \end{bmatrix}$$

$$[T_{in,2}] = \begin{bmatrix} 0 & 0 \\ 0 & \dfrac{L_{out}C_{out}}{\Delta t^2} \end{bmatrix}$$

Similarly, the output node of the transmission line is modeled by

$$\begin{bmatrix} V_{in}(t) \\ J_{in}(t) \end{bmatrix} = \sum_{k\geq 0} [T_{out,k}] \begin{bmatrix} V(z,t-k\Delta t) \\ J(z,t-k\Delta t) \end{bmatrix} \qquad [6.6]$$

Thus, we get five matrix equations where only the functions $V_{out}(t)$ (voltage source) or $J_{out}(t)$ (current source) and $J_{in} = 0$ (or $V_{in} = 0$) are known as boundary conditions. This is shown in Figures 6.3 and 6.4 depicting the case of a resistive load in parallel with a capacitor with $R_{out} = L_{out} = C_{out} = L_{in} = 0$.

6.1.6. *Complex distributed and lumped parameter networks approximation*

In realistic applications, the lumped networks modeling each node of a complex network contain elements having lumped parameters depending on frequency such as inductance L and capacitance C. Otherwise, everything is similar to that presented previously in the discussion about the complex network processed according to the "direct" method.

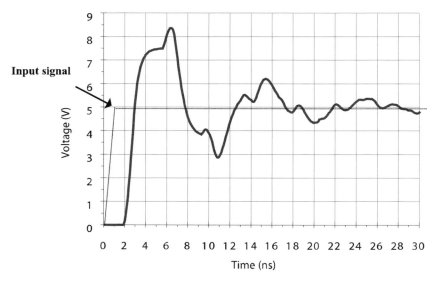

Figure 6.3. *Single line with $R > Z_{c0}$ and $C_{in} = 5\ pF$*

By means of the discretization process, the *state equation* becomes

$$[A]\left[X\left(t - \frac{[z]}{v}\right)\right]$$

$$= \left[Y\left(t - \frac{[z_0]}{v}\right)\right] + \sum_{k=0}^{k=K}\left\{[C_k]\left[X\left(t - \frac{[z]}{v} - k\Delta t\right)\right] + [B_k]\left[X\left(t - \frac{2[z]}{v} - k\Delta t\right)\right]\right\} \quad [6.7]$$

where $[A]$, $[C]$, and $[B_k]$ are matrices having the same structure as previously shown in section 5.3.4 in respect of the objective defined in section 2.3.3. Nevertheless, the example in Figure 6.5 shows the complexity of the matrix structure is greater than in the case of the direct method in the time domain that strictly depends on the accuracy of the lumped element modeling.

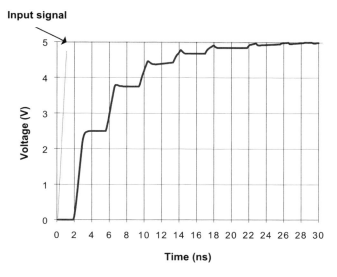

Figure 6.4. *Single line with $R < Z_{c0}$ and $C_{in} = 5\ pF$*

There are new vectors involved in equation [6.7]:

$$\left[X\left(t - \frac{[z]}{v} - k\Delta t\right)\right]$$

is the already known voltage and current vector at a past time due to the use of the finite difference approximation of the nodal operational matrices modeling the frequency-dependent lumped elements, and

$$\left[X\left(t - \frac{2[z]}{v} - k\Delta t\right)\right]$$

is the voltage and current vector already known because it arises from reflections with lumped elements depending on frequency.

178   Electromagnetism and Interconnections

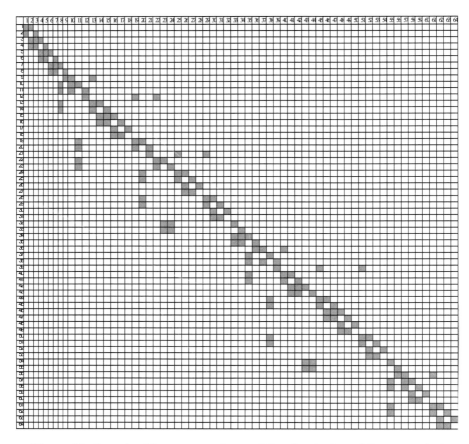

Figure 6.5. *Example of the structure of matrix [A] tied to the finite difference method (gray: non-zero elements; white: zero elements)*

Then, equation [6.7] is solved with respect to the unknown vector by means of the Gauss algorithm which has to be used only once in the case of frequency-dependent linear lumped elements and as many times as necessary in the case of frequency-dependent nonlinear lumped elements.

How can we estimate the computation requirements related to the above process?

In order to answer this, let L be the number of transmission line segments. Choosing the Gauss algorithm needs $(1/3)(4L)^3$ basic operations like additions and multiplications.

Furthermore, it is necessary to store $(K+2)[(4L)^2 + 4L]$ numbers in the system memory, K being the order of the linear models used for the simulation of the lumped elements.

Appendix D shows a modified Gauss–Seidel algorithm which is a recursive method of solving [6.7] by saving a large amount of storage capacity.

6.2. Matrix velocity operator interpolation method

6.2.1. *Difficulties set by the compounded matrix functions*

Let the column matrix $[u]$ be defined by its transposed row matrix

$$[u]^T = \left[[e_x][e_y][[e][h_x][h_y][h]] \right]$$

We obtain the following first-order partial derivative matrix equation having three constant matrix coefficients which do not commute:

$$M\frac{\partial u}{\partial t} + N\frac{\partial u}{\partial z} + Ru = 0$$

We would like to solve the straight heterogenous media equation by means of the time-domain "propagator" or "propagation operator" introduced for homogenous lossless lines. Before this, we have to focus on the "compounded matrix functions" that are required.

A matrix function defined on the set of real numbers is any mapping from this set to the set of matrices, which are, in our application, square with infinite dimension. It is known that any sum of them defines a new matrix and also their product with any scalar function: so, they belong to a vector space. The derivative of them versus the real variable on which they depend is a linear operator working in this vector space. This leads to a new matrix function belonging to the vector subset of the differentiable matrix functions.

Moreover, the product of these matrices is a non-commutative operation defining an algebraic structure to their set. The classical differentiation rule of any matrix product is always valid in so far as the order of the matrices in the product is respected. Up to now everything is straightforward.

The difficulties arise with the compounded matrix functions needed for solving first-order differential and partial differential matrix equations having matrix coefficients. The situation is rigorously depicted as follows. Let f be a mapping from the set of infinite dimension matrices like A to this set. Since A is dependent on the scalar variable t, we get a new matrix function defined on the set of real numbers called a *compounded matrix function*:

$$f \circ A(t) = f[A(t)]$$

where f is called an *isotropic mapping* since, whatever the invertible infinite dimension matrix P, the following relation is true:

$$f(P^{-1}AP) = P^{-1}f(A)P$$

In this case, $f(A)$ can be expanded into a Taylor series having scalar coefficients and it commutes with its differential versus A, denoted $f'(A)$. Because the dimension of A is assumed to be infinite, the previous expansion is related to Dunford's operational calculus [DAU3] where the linear operators and their associated infinite dimension matrices are processed like classical (commutative) algebraic numbers. Let us introduce the following compounded matrix function:

$$f' \circ A(t) = f'[A(t)]$$

Now, is the classical rule of differentiating the compounded matrix functions always true?

Let us recall this rule as follows:

$$\frac{d}{dt}[f' \circ A(t)] = A'(t)[f' \circ A(t)] \quad \text{or} \quad \frac{d}{dt}[f(A)] = A'f'(A)$$

This rule has to be true for any other function; then dealing with $Af(A)$, we obtain directly

$$\frac{d}{dt}[Af(A)] = A'f(A) + AA'f'(A)$$

while the application of the rule requires

$$\frac{d}{dt}[Af(A)] = A'[f(A) + Af'(A)]$$

The two last equalities are similar whatever f, if and only if $AA' = A'A$.

The classical rule of *differentiating the compounded matrix function* $f[A(t)]$ is true if and only if the matrix function $A(t)$ commutes with its derivative. In this case, the *compounded matrix function* is called *isotropic*.

Here is the main limitation of the application of classical matrix functions for solving the matrix differential equation

$$\frac{d[u]}{dt} = A(t)[u]$$

where the matrix A is neither constant nor even diagonal.

Note that differentiating the relation $AA' = A'A$, we obtain

$$AA'' = A''A$$

Multiplying this on both sides by A^{-1} gives

$$A''A^{-1} = A^{-1}A''$$

and

$$A''A^{-1}AA' = A''A' = A'AA^{-1}A'' = A'A''$$

Step by step, it is shown that all the differentials of any order commute.

So, all the matrix coefficients of the Taylor expansion of $A(t)$ versus t have to commute in order to make any compounded matrix function $f[A(t)]$ isotropic. For instance, $f(A + Bt)$ is isotropic if and only if A and B commute.

Among the existing compounded matrix functions, those like $f[Bg(t)]$, where B is a constant matrix and g a scalar function, are always isotropic. So is $f(Bt)$.

6.2.2. *Matrix velocity operator of stratified heterogenous media*

Let the column matrix $[u]$ be defined by its transposed row matrix

$$[u]^T = \big[[e_x][e_y][[e][h_x][h_y][h]\big]$$

We obtain the following first-order partial derivative matrix equation having three constant matrix coefficients which do not commute:

$$M\frac{\partial u}{\partial t} + N\frac{\partial u}{\partial z} + Ru = 0$$

We propose in the following a process of reducing the three constant coefficients which do not commute to a single but to a time-varying coefficient.

M is invertible; it is possible to set

$$u = \exp(-M^{-1}Rt)v \qquad [6.8]$$

from which

$$\frac{\partial u}{\partial t} = \exp(-M^{-1}Rt)\left(\frac{\partial v}{\partial t} - M^{-1}Rv\right)$$

$$M\exp(-M^{-1}Rt)\frac{\partial v}{\partial t} - M\exp(-M^{-1}Rt)M^{-1}Rv + N\exp(-M^{-1}Rt)\frac{\partial v}{\partial z}$$
$$+ R\exp(-M^{-1}Rt)v = 0$$

The matrix exponential $\exp(-M^{-1}Rt)$ commutes with the matrix $M^{-1}R$; then

$$M\exp(-M^{-1}Rt)M^{-1}R = M(M^{-1}R)\exp(-M^{-1}Rt) = R\exp(-M^{-1}Rt)$$

So, we obtain

$$M\exp(-M^{-1}Rt)\frac{\partial v}{\partial t} + N\exp(-M^{-1}Rt)\frac{\partial v}{\partial z} = 0$$

R is not invertible; therefore $M^{-1}R$ is not invertible.

However,

$$\exp(-M^{-1}Rt) = I - M^{-1}Rt + \frac{1}{2}(M^{-1}Rt)^2 t^2 + \cdots$$

is invertible, and we can write

$$\frac{\partial v}{\partial t} + \underbrace{\left[M\exp(-M^{-1}Rt)\right]^{-1}}_{\exp(M^{-1}Rt)M^{-1}} N\exp(-M^{-1}Rt)\frac{\partial v}{\partial z} = 0$$

$$\underbrace{\exp(M^{-1}Rt)M^{-1}N\exp(-M^{-1}Rt)}_{} = V$$

This infinite dimension matrix V defines the *matrix velocity operator* (not invertible).

In what follows, we focus our attention on the following matrix functions:

$$V(A,B,t) = \exp(At)B\exp(-At)$$

The matrix velocity operator is a generalization of the classical scalar velocity to straight heterogenous stratified media. Its introduction is justified by the next process of solving the straight heterogenous media equation:

$$\frac{\partial v}{\partial t} + V\frac{\partial v}{\partial z} = 0 \qquad [6.9]$$

This is a *matrix drift equation* model [CAR].

6.2.3. *Matrix velocity operator interpolation method for the matrix drift equation*

It is always possible to consider the primitive function $Z(t)$ of the matrix function $V(t)$ so that we can write

$$\frac{\partial v}{\partial t} + Z'(t)\frac{\partial v}{\partial z} = 0$$

We would like to get the solution as follows: $v(z, t) = f[Z(t)]$, which is a compounded matrix function. According to the previous demonstration, it is necessary to account for isotropy.

Because of this, we need to have a look at some properties of the matrix function $V(t)$.

Let $V(A,B,t) = \exp(At)B\exp(-At)$; of course $V(A,B,0) = B$, and we obtain the following properties.

Property of translation:

$$V(A,B,t-t_0) = \exp(-At_0)V(A,B,t)\exp(At_0)$$

Property of differentiation:

$$\frac{\partial}{\partial t} V(A,B,t) = \exp(-At)(AB - BA)\exp(At) = V(A, AB - BA, t)$$

It can be seen that the differential is zero since A and B commute, then V is constant.

This yields the Taylor expansion of the velocity matrix operator:

$$V(A,B,t) = \sum_{n=0}^{\infty} \left(\frac{\partial^n V}{\partial t^n}\right)_{t=t_0} \frac{(t-t_0)^n}{n!}$$

with

$$\left(\frac{\partial^n V}{\partial t^n}\right)_{t=0} = A\left(\frac{\partial^{n-1} V}{\partial t^{n-1}}\right)_{t=0} - \left(\frac{\partial^{n-1} V}{\partial t^{n-1}}\right)_{t=0} A, \quad n > 1$$

and

$$\left(\frac{\partial^0 V}{\partial t^0}\right)_{t=0} = B$$

So, the integration of $V(t)$ gives $Z(t)$ in a Taylor expansion form which is well-suited for defining an efficient approximation method based on the use of isotropy.

The function $f[Z(t)]$ is an available isotropic solution of the equation

$$\frac{\partial v}{\partial t} + Z'(t) \frac{\partial v}{\partial z} = 0$$

since $Z(t)$ is equal to Bt, that is never rigorously but only approximately true within a short period of time. We now evaluate by means of the Taylor expansion:

$$Z(A,B,t) - Z(A,B,t_0) - B(t - t_0) = \sum_{n=1}^{\infty} \left(\frac{\partial^n V}{\partial t^n}\right)_{t=t_0} \frac{(t-t_0)^n}{(n+1)!}$$

Applying the properties of the norms of matrices, we deal with the following recursive inequalities:

$$\left\|\left(\frac{\partial^n V}{\partial t^n}\right)_{t=0}\right\| \le \left\|A\left(\frac{\partial^{n-1} V}{\partial t^{n-1}}\right)_{t=0}\right\| + \left\|\left(\frac{\partial^{n-1} V}{\partial t^{n-1}}\right)_{t=0} A\right\| \le 2\|A\|\left\|\left(\frac{\partial^{n-1} V}{\partial t^{n-1}}\right)_{t=0}\right\|$$

$$\le (2\|A\|)^{n-1}\|AB - BA\|, \quad n \ge 1$$

from which, since $t > t_0$, we obtain

$$\|Z(A,B,t) - Z(A,B,t_0) - B(t-t_0)\| \le \sum_{n=1}^{\infty}\left\|\left(\frac{\partial^n V}{\partial t^n}\right)_{t=t_0}\right\|\frac{(t-t_0)^{n+1}}{(n+1)!} \le \|A\|\|B\|(t-t_0)^2 e^{\|A\|(t-t_0)}$$

As $n(n+1) \ge 2$ when $n \ge 1$, we can write

$$\frac{(t-t_0)^{n+1}}{(n+1)!} = \frac{(t-t_0)^2}{n(n+1)!}\frac{(t-t_0)^{n-1}}{(n-1)!} \le \frac{(t-t_0)^2}{2}\frac{(t-t_0)^{n-1}}{(n-1)!}$$

which allows the following majoration:

$$\|Z(A,B,t) - Z(A,B,t_0) - B(t-t_0)\| \le \|AB - BA\|\sum_{n=1}^{\infty}\|2A\|^{n-1}\frac{(t-t_0)^2}{2}\frac{(t-t_0)^{n-1}}{(n-1)!}$$

$$= \|AB - BA\|\frac{(t-t_0)^2}{2}e^{2\|A\|(t-t_0)}$$

Setting this smaller than any given number yields a value t_1 of how many times the linear variation of Z versus time is valid. During $[0, t_1]$, this linearity leads to a solution given by a compounded matrix function in so far as $Z(A, B, t_0)$ commutes with B that is never the case except when it is zero, which we now assume. $Z(A, B, t_0)$ being zero, we have to find the solution of the straight heterogenous media equation written in the form

$$\frac{\partial v}{\partial t} + \frac{\partial}{\partial z}[V(t)v] = 0 \qquad [6.10]$$

the so-called *drift equation* corresponding to a known boundary condition $v(0, t)$ at $z = 0$ or to an initial condition $v(z, 0)$ at $t = 0$.

Note the most general boundary condition $v(0, t)$ can be expressed as a sum of products of isotropic compounded matrix functions like $f(Bt)$ by constant vectors like v_0. In this case, the estimate solution of the drift equation corresponding to the boundary conditions $v(0, t) = f(Bt)v_0$ becomes

$$v(z,t) = f(Bt - zI)v_0$$

where the constant matrix B has the spectral expansion

$$B - \sum_i \lambda_i P_i$$

in which the matrices P_i being the projectors of B and f are defined by

$$f(Bt - zI) = \sum_{\lambda_i t - z \geq 0} f(\lambda_i t - z) P_i \qquad [6.11]$$

which is valid during $[0, t_1]$, where t_1 is such that

$$\|AB - BA\| \frac{(t-t_0)^2}{2} e^{2\|A\|(t-t_0)} < \varepsilon$$

After time t_1, we process a translation of time with t_1 as a new origin. The property of translation gives

$$V(A, B, t) = \exp(At_1) V(A, B, t - t_1) \exp(-At_1)$$

This means matrix B is replaced by $B_1 = \exp(At_1) B \exp(-At_1)$ during $[t_1, t_2]$ with

$$\|AB - BA\| \frac{(t-t_0)^2}{2} e^{2\|A\|(t_2 - t_1)} < \varepsilon$$

Then $v(z,t) = f[B_1(t - t_1) - zI]v_1$ with $v_1 = f(Bt_1 - zI)v_0$, and so on.

Therefore, at a time t belonging to $[tn, t_{n+1}]$ where $t_n = n\tau$ with

$$\|AB - BA\| \frac{\tau^2}{2} e^{2\|A\|\tau} < \varepsilon$$

the solution is

$$\begin{aligned} v(z,t) &= f[B_n(t - t_n) - zI] f[B_{n-1}(t_n - t_{n-1}) - zI] \\ &\quad \ldots f[B_1(t_2 - t_1) - zI] f[B(t_1 - t_0) - zI] v_0 \end{aligned} \qquad [6.12]$$

the order of the above matrix factors having been respected. This achieves the matrix velocity operator interpolation method.

In the case of known initial conditions made up of the sum of matrix functions like

$$v(z,0) = f(zI)v_0$$

we obtain

$$v(z,t) = f(zI - Bt)v_0$$

with

$$f(Bt - zI) = \sum_{\lambda_i t - z \geq 0} f(\lambda_i t - z) P_i$$

during $[0, t_1]$, and so on as above.

6.3. Conclusion

The previously mentioned matrix state equation, obtained by the direct method, has been completed by the finite difference modeling of lumped elements in the time domain.

The generalized transmission line matrix equation of heterogenous stratified media has been processed by a discretization method based on the use of a "velocity matrix operator" linear operator.

While the storage memory space required by these time-domain discretization methods remains similar to that of the direct time-domain methods, the computation time of waveform amplitude at any time increases with the number of time steps which have to be at the most a tenth of the transition times of digital signals.

Chapter 7

Frequency Methods

The transfer matrix is defined in the Laplace domain for any set of lossy (both dielectrics and conductors) coupled transmission lines having the same crosstalk length. Then, the spectral theory of matrices is applied to the slow signals assuming a constant lineic resistance and to the high-speed signals using the Foucault modal currents previously developed. The multiple reflections in the Laplace domain are studied by means of the *scattering* matrix. The development of irrational functions into an infinitely continued fraction is applied in order to obtain the lumped circuit equivalent to any lossy characteristic impedance, thus showing the way for an approximated impedance matching.

The theory of many-valued analytical functions is needed for coming back to the time domain, so leading to the convolution methods for lossy coupled line modeling. Asymptotic approximation of a Dirac pulse propagation is applied to *skin effect* modeling in the time domain.

We deal with the pure frequency analysis based on the Fourier transform and the Shannon interpolating formula which are applied to digital waveforms and their spectrum range estimate. This is achieved by the *matrix state equation* in the frequency domain for complex lossy network modeling with the estimation of its computation and memory requirements.

7.1. Laplace transform method for lossy transmission lines

7.1.1. *Transfer matrix in the Laplace domain*

7.1.1.1. *Modeling lossy conductors with a common continuous lineic resistor*

7.1.1.1.1. The common continuous resistor approximation

As the propagation mode is quasi-TEM, every electromagnetic wave has the same velocity v, and it is possible to define, in this case, capacitance and inductance matrices per unit length, $[C]$ and $[L]$ respectively, to which the characteristic impedance $[Z_{c0}] = (1/v)[C]^{-1}$ [5.17] is related.

Then, the electromagnetic influence matrix between the current and voltage vectors is defined in terms of the matrix block already met for lossless transmission lines:

$$v\begin{bmatrix}[0] & [L]\\ [C] & [0]\end{bmatrix} = [M]\begin{bmatrix}[0] & [Z_{c0}]\\ [Z_{c0}]^{-1} & [0]\end{bmatrix} \quad [7.1]$$

All the losses in both conductors and dielectrics are accounted for by the following matrix block:

$$[N] = \begin{bmatrix}[0] & [R]\\ [G] & [0]\end{bmatrix} v \quad [7.2]$$

$[G]$ being the conductance matrix per unit length which is related to the characteristic impedance matrix by means of the relation

$$[G] = \frac{\sigma}{\varepsilon v}[Z_{c0}]^{-1}$$

and $[R]$ being the resistor matrix per unit length defined in the case of continuous steady state of electrical currents inside lossy conductors.

Let us consider the case of "slowly variable" waveforms that correspond to clock frequency up to a few megahertz, so allowing the continuous state assumption. In the applications to interconnections modeling within multilayered substrates used in electronics, all the conductors have equal cross-sections and, therefore, have the same continuous lineic resistor r. Thus, $[R] = r[1]$, where $[1]$ is the unit matrix, so the matrices $[Z_{c0}]$ or $[Z_{c0}]^{-1}$ and $[R]$ have a commutative product.

This last property is very useful for solving the matrix equation of coupled lossy transmission lines having the same length. This equation can be written as follows:

$$v\frac{\partial}{\partial z}\begin{bmatrix}[V]\\[J]\end{bmatrix}+[M]\frac{\partial}{\partial t}\begin{bmatrix}[V]\\[J]\end{bmatrix}+[N]\begin{bmatrix}[V]\\[J]\end{bmatrix}=\begin{bmatrix}[0]\\[0]\end{bmatrix} \quad [7.3]$$

[V] and [J] being the voltage and current vectors and [M] being the electromagnetic mutual influence matrix.

7.1.1.1.2. Spectral representation of the transfer matrix

Let us demonstrate the method of classical analysis in the frequency domain. In order to get both the transient and the steady states at the same time, we have to use the Laplace transform of which the operator is denoted as L, the result of the transformation of a function f being noted as $\hat{f}$, and p being the symbolic variable which is related to the "complex frequency" ω by means of $p = j\omega$. According to these definitions, we write

$$\begin{bmatrix}[\hat{V}(z,p)]\\ \hat{J}(z,p)\end{bmatrix}=L\begin{bmatrix}[V(z,t)]\\ J(z,t)\end{bmatrix}=\int_0^\infty e^{-pt}\begin{bmatrix}[V(z,t)]\\ J(z,t)\end{bmatrix}dt \quad [7.4]$$

Remembering that $L[f'(t)] = p\hat{f}(p) - f(0)$, we obtain

$$v\frac{\partial}{\partial z}\begin{bmatrix}[\hat{V}]\\[\hat{J}]\end{bmatrix}+\{[M]p+[N]\}\begin{bmatrix}[\hat{V}]\\[\hat{J}]\end{bmatrix}=\begin{bmatrix}[V(z,0)]\\[J(z,0)]\end{bmatrix}$$

It is possible to find the solution of the matrix by means of an exponential matrix which defines the *transfer matrix of the set of coupled lines*:

$$\begin{bmatrix}[\hat{V}(z,p)]\\[\hat{J}(z,p)]\end{bmatrix}=\underbrace{\exp\left\{-\{[M]p+[N]\}\frac{z}{v}\right\}}_{\text{transfer matrix}}\begin{bmatrix}[\hat{V}(0,p)]\\[\hat{J}(0,p)]\end{bmatrix} \quad [7.5]$$

However, matrices [M] and [N] do not commute, so the above exponential of the sum of two matrices cannot be equal to the product of the exponential of each matrix separately.

We will apply in what follows the theory of the *spectral expansion of matrix functions* [LAS3]. Let [A] be a square matrix having all its n eigenvalues λ_m different; it has the following spectral expansion:

$$[A] = \sum_{m=1}^{m=n} \lambda_m [P_m]$$

where $[P_m]$ are "projectors" which are matrices meeting the following properties that have been already mentioned.

These properties are

$$\begin{cases} [P_m]^2 = [P_m] \\ [P_m][P_1] = [0] \quad \text{if} \quad m \neq 1 \end{cases}$$

According to the *Duncan-Sylvester formula*:

$$P_m = \prod_{1 \neq m} \frac{[A] - \lambda_1 [1]}{\lambda_m - \lambda_1}$$

Then, it can be shown classically:

$$f([A]) = \sum_{m=1}^{m=n} f(\lambda_m)[P_m]$$

where

$$[A] = \frac{1}{v}\{[M]p + [N]\} = \begin{bmatrix} [0] & \frac{p}{v}[Z_{c0}] + [R] \\ \frac{p}{v}[Z_{c0}]^{-1} + [G] & [0] \end{bmatrix}$$

and $f(\lambda_m) = e^{-\lambda_m z}$.

7.1.1.1.3. Modal complex parameters of propagation

The equation of the eigenvectors of $[A]$ is

$$\begin{bmatrix} -\lambda[1] & [Z_{c0}]\frac{p}{v} + [R] \\ [Z_{c0}]^{-1}\frac{p}{v} + [G] & -\lambda[1] \end{bmatrix} \begin{bmatrix} [V] \\ [J] \end{bmatrix} = \begin{bmatrix} [0] \\ [0] \end{bmatrix} \quad [7.6]$$

which gives

$$\begin{cases} \lambda[V] = \left\{ [Z_{c0}]\dfrac{p}{v} + [R] \right\}[J] \\ \lambda[J] = \left\{ [Z_{c0}]^{-1}\dfrac{p}{v} + [G] \right\}[V] \end{cases}$$

Using the relation $[Z_{c0}] = (1/v)[C]^{-1}$, which we have already encountered, then separating the vectors $[V]$ and $[J]$, leads to the eigenvector equations of the matrices $\{[R][Z_{c0}]^{-1}\}$ and $\{[Z_{c0}]^{-1}[R]\}$, respectively:

$$\left\{ [R][Z_{c0}]^{-1} - \dfrac{v^2\lambda^2 - p^2 - \dfrac{\sigma}{\varepsilon}p}{\left(p + \dfrac{\sigma}{\varepsilon}\right)v}[1] \right\}[V] = [0]$$

$$\left\{ [Z_{c0}]^{-1}[R] - \dfrac{v^2\lambda^2 - p^2 - \dfrac{\sigma}{\varepsilon}p}{\left(p + \dfrac{\sigma}{\varepsilon}\right)v}[1] \right\}[J] = [0]$$

Since the matrices $[R]$ and $[Z_{c0}]^{-1}$ have a commutative product (because of the common lineic resistor of all the conductors), the above equations are strictly similar, and there are n couples of opposite eigenvalues $\lambda = \pm\lambda_m$ for matrix $[A]$.

Remembering that $[Z_{c0}] = (1/v)[C]^{-1}$, these eigenvalues are easily related to those c_m of the capacitor matrix of which the eigenvalue equation is $\{[C] - c[1]\}[V] = 0$.

Indeed, $[Z_{c0}]^{-1}[R] = rv[C]$, r being the *common resistance per unit length* of all the conductors.

Therefore

$$c_m = \dfrac{v^2\lambda_m^2 - p^2 - \dfrac{\sigma}{\varepsilon}p}{\left(p + \dfrac{\sigma}{\varepsilon}\right)rv^2}$$

and

$$\lambda_m = \pm \frac{1}{v}\sqrt{\left(p+\frac{\sigma}{\varepsilon}\right)(p+rv^2 c_m)} \qquad [7.7]$$

The *complex parameter of propagation* is defined by

$$\gamma_m = |\lambda_m| = \sqrt{\left(p+\frac{\sigma}{\varepsilon}\right)(p+rv^2 c_m)}$$

7.1.1.1.4. Modal characteristic and crosstalk impedance matrices

Accounting for the couple of opposite eigenvalues, the transfer matrix becomes

$$[T(p)] = \exp\{-([M]p+[N])\} = \sum_{\lambda_m>0}[P'_m]e^{-\gamma_m z} + \sum_{\lambda_m<0}[P''_m]e^{-\gamma_m z} \qquad [7.8]$$

with

$$[P'_m] = \frac{[A]-(-\gamma_m)[1]}{\gamma_m-(-\gamma_m)} \prod_{l\neq m}^{1\leq n}\frac{[A]-\gamma_1[1]}{\gamma_m-\gamma_1}\prod_{l\neq m}^{1\leq n}\frac{[A]+\gamma_1[1]}{\gamma_m+\gamma_1} = \frac{[A]+\gamma_m[1]}{2\gamma_m}\prod_{l\neq m}^{1\leq n}\frac{[A]^2-\gamma_1^2[1]}{\gamma_m^2-\gamma_1^2}$$

Similarly, we obtain

$$[P''_m] = \frac{[A]-\gamma_m[1]}{-2\gamma_m}\prod_{l\neq m}^{1\leq n}\frac{[A]^2-\gamma_1^2[1]}{\gamma_m^2-\gamma_1^2}$$

From these last relations, we obtain

$$[P'_m]-[P''_m] = \prod_{l\neq m}^{1\leq n}\frac{[A]^2-\gamma_1^2[1]}{\gamma_m^2-\gamma_1^2}$$

Thus

$$[P'_m] = \frac{[A]+\gamma_m[1]}{2\gamma_m}\{[P'_m]-[P''_m]\}$$

and

$$[P_m''] = \frac{[A] - \gamma_m[1]}{-2\gamma_m}\{[P_m'] - [P_m'']\}$$

Therefore, a new expression of the transfer matrix more classically suited for transmission line analysis can be obtained by means of the above relations:

$$[T(p)] = \sum_m \frac{[A] + \gamma_m[1]}{2\gamma_m}\{[P_m'] - [P_m'']\}e^{-\gamma_m z} + \sum_m \frac{[A] - \gamma_m[1]}{-2\gamma_m}\{[P_m'] - [P_m'']\}e^{-\gamma_m z}$$

$$[T(p)] = \sum_m \left\{[1]\text{ch}(\gamma_m z) - [A]\frac{\text{sh}(\gamma_m z)}{\gamma_m}\right\}\{[P_m'] - [P_m'']\}$$

Accounting for $[Z_{c0}] = (1/v)[C]^{-1}$ so

$$[R] = rvc_m[Z_{c0}] + rv\{[C] - c_m[1]\}[Z_{c0}] \quad \text{and} \quad [G] = \frac{\sigma}{\varepsilon v}[Z_{c0}]^{-1}$$

we obtain the next matrix defining new sub-matrices as follows:

$$\frac{[A]}{\gamma_m} = \begin{bmatrix} [0] & \dfrac{p + rv^2 c_m + rv^2\{[C] - c_m[1]\}}{\sqrt{\left(p + \dfrac{\sigma}{\varepsilon}\right)(p + rv^2 c_m)}}[Z_{c0}] \\ \dfrac{\left(p + \dfrac{\sigma}{\varepsilon}\right)[Z_{c0}]}{\sqrt{\left(p + \dfrac{\sigma}{\varepsilon}\right)(p + rv^2 c_m)}} & [0] \end{bmatrix}$$

$$= \begin{bmatrix} [0] & [\hat{Z}_{cm}(p)] + [\hat{Z}_{dm}(p)] \\ [\hat{Z}_{cm}(p)]^{-1} & [0] \end{bmatrix}$$

where we have the *modal characteristic impedance matrix* of the coupled lines:

$$[\hat{Z}_{cm}(p)] = \zeta_{cm}(p)[Z_{c0}(p)] \quad \text{with} \quad \zeta_{cm}(p) = \sqrt{\frac{p + rv^2 c_m}{p + \dfrac{\sigma}{\varepsilon}}}$$

and the *modal crosstalk impedance matrix* of the coupling capacitors between lines:

196   Electromagnetism and Interconnections

$$[\hat{Z}_{dm}(p)] = \frac{rv^2}{\sqrt{\left(p+\frac{\sigma}{\varepsilon}\right)(p+rv^2 c_m)}}\{[C]-c_m[1]\}[Z_{c0}]$$

which is zero since the capacitor matrix is diagonal. Then, we obtain the final form of the transfer matrix:

$$[T(p)] = \sum_m [T_m(p)]\{[P'_m]-[P''_m]\}$$

with

$$[T_m(z,p)] = \begin{bmatrix} [1]\mathrm{ch}(\gamma_m z) & -\{[\hat{Z}_{cm}(p)]+[\hat{Z}_{dm}(p)]\}\mathrm{sh}(\gamma_m z) \\ -[\hat{Z}_{cm}]^{-1}\mathrm{sh}(\gamma_m z) & [1]\mathrm{ch}(\gamma_m z) \end{bmatrix}$$

which defines the *modal transfer matrix* in the case of a *common continuous lineic resistor*.

7.1.1.2. *Modeling lossy conductors rigorously with Foucault's modal lineic resistors*

In the case of very high-speed signals needing the use of the Foucault currents modal representation, there are only two different eigenvalues solution of equation [7.6], and the Duncan–Sylvester approach is no longer applicable. Recalling the relation $[q] = [C][V]$ for each mode having the index i, we deal with the following matrix equation:

$$\frac{\partial}{\partial z}\begin{bmatrix} q_i \\ j_i \end{bmatrix} + \begin{bmatrix} 0 & \mu_0\varepsilon_d - \frac{\varepsilon_c^2\beta_i^2}{\sigma_c^2} \\ 1 & 0 \end{bmatrix}\frac{\partial}{\partial t}\begin{bmatrix} q_i \\ j_i \end{bmatrix} + \begin{bmatrix} 0 & \beta_i^2\frac{\varepsilon_c}{\sigma_c} \\ \frac{\sigma_d}{\varepsilon_d} & 0 \end{bmatrix}\begin{bmatrix} q_i \\ j_i \end{bmatrix} = \begin{bmatrix} 0 \\ 0 \end{bmatrix} \quad [7.9]$$

In the Laplace domain, this becomes

$$\frac{\partial}{\partial z}\begin{bmatrix} \hat{q} \\ \hat{j} \end{bmatrix} + \underbrace{\begin{bmatrix} 0 & \left(\mu_0\varepsilon_d - \frac{\varepsilon_c^2\beta_i^2}{\sigma_c^2}\right)p + \beta_i^2\frac{\varepsilon_c}{\sigma_c} \\ p+\frac{\sigma_d}{\varepsilon_d} & 0 \end{bmatrix}}_{[A_i]}\begin{bmatrix} \hat{q} \\ \hat{j} \end{bmatrix} = \begin{bmatrix} 0 \\ 0 \end{bmatrix} \quad [7.10]$$

so we get a square matrix $[A_i]$ which is of size 2×2 and different eigenvalues λ_{i1} and λ_{i2}:

$$\lambda_{i1} = -\lambda_{i2} = \sqrt{\left(p + \frac{\sigma_d}{\varepsilon_d}\right)\left[\left(\mu_0 \varepsilon_d - \frac{\varepsilon_c^2 \beta_i^2}{\sigma_{ci}^2}\right) p + \beta_i^2 \frac{\varepsilon_c}{\sigma_{ci}}\right]} \quad [7.11]$$

and the complex parameter of propagation related to mode i defined by

$$\gamma_i = \lambda_{i1} \quad [7.12]$$

Then, the matrix exponential can be expressed by means of the Duncan–Sylvester formula:

$$\exp\{-[A_i]z\} = e^{-\lambda_{i1}z} \frac{[A_i] - \lambda_{i2}[1]}{\lambda_{i1} - \lambda_{i2}} + e^{-\lambda_{i2}z} \frac{[A_i] - \lambda_{i12}[1]}{\lambda_{i2} - \lambda_{i1}}$$

which gives the transfer matrix

$$[1]\mathrm{ch}(\lambda_{i1}z) - [A_i]\frac{\mathrm{sh}(\lambda_{i1}z)}{\lambda_{i1}} \quad [7.13]$$

within which we can highlight the parameter ζ_{ci} defined by

$$\zeta_{ci}(p) = \sqrt{\frac{\left(\mu_0 \varepsilon_d - \frac{\varepsilon_c^2 \beta_i^2}{\sigma_c^2}\right) p + \frac{\varepsilon_d}{\sigma_c} \beta_i^2}{p + \frac{\sigma_d}{\varepsilon_c}}} \quad [7.14]$$

so that we obtain the very classical transfer matrix:

$$\text{transfer matrix} = \begin{bmatrix} \mathrm{ch}(\gamma_i z) & -\zeta_{ci} \frac{\mathrm{sh}(\gamma_i z)}{v} \\ -v \frac{\mathrm{sh}(\gamma_i z)}{\zeta_{ci}} & \mathrm{ch}(\gamma_i z) \end{bmatrix} \quad [7.15]$$

At the level of the whole set of coupled lossy lines having the same length, the usual matrix blocks are applied with the relation $[q_i] = [C][V_i]$, thus obtaining the modal transfer matrix:

$$\begin{bmatrix} [1]\text{ch}(\gamma_i z) & -\dfrac{\zeta_{ci}}{v}\text{sh}(\gamma_i z)[C]^{-1} \\ -\dfrac{v}{\zeta_{ci}}\text{sh}(\gamma_i z)[C] & [1]\text{ch}(\gamma_i z) \end{bmatrix} \qquad [7.16]$$

Then, the *characteristic impedance matrix* in the Laplace domain becomes

$$[\hat{Z}_{ci}(p)] = \zeta_{ci}[Z_{c0}] \qquad [7.17]$$

and the *transfer matrix* becomes

$$[T_i(z,p)] = \begin{bmatrix} [1]\text{ch}(\gamma_i z) & -[\hat{Z}_{ci}]\text{sh}(\gamma_i z) \\ -[\hat{Z}_{ci}]^{-1}\text{sh}(\gamma_i z) & [1]\text{ch}(\gamma_i z) \end{bmatrix} \qquad [7.18]$$

To finish, each mode satisfies the following *matrix transfer equation*:

$$\begin{bmatrix} [\hat{V}_i(z,p)] \\ [\hat{J}_i(z,p)] \end{bmatrix} = [T_i(z,p)] \begin{bmatrix} [\hat{V}_i(0,p)] \\ [\hat{J}_i(0,p)] \end{bmatrix} \qquad [7.19]$$

7.1.2. Transfer impedance matrix, impedance matching, scattering matrix

7.1.2.1. *Foucault's modal transfer impedance matrix*

Let us consider any linear load network characterized by the impedance matrix $[\hat{Z}(p)]$ in the Laplace domain at the line terminals. We have the relation

$$[\hat{V}_i(z,p)] = [\hat{Z}(p)][\hat{J}_i(z,p)] \qquad [7.20]$$

Applying the transfer relation [7.19] and the expression of the transfer matrix [7.18], we obtain the two following equations from relation [7.20]:

$$\begin{aligned} [\hat{V}_i(z,p)] &= \text{ch}(\gamma_i z)[\hat{V}_i(0,p)] - [\hat{Z}_{ci}]\text{sh}(\gamma_i z)[\hat{J}_i(0,p)] \\ \left[[Z(p)]^{-1}[\hat{V}_i(z,p)]\right] &= -[\hat{Z}_{ci}]^{-1}\text{sh}(\gamma_i z)[\hat{V}_i(0,p)] + \text{ch}(\gamma_i z)[\hat{J}_i(0,p)] \end{aligned} \qquad [7.21]$$

Eliminating $[\hat{J}_i(0,p)]$ between the previous relations, we obtain

$$[\hat{V}_i(z,p)] = \left\{[1]\text{ch}(\gamma_i z) + [\hat{Z}_{ci}][\hat{Z}(p)]^{-1}\text{sh}(\gamma_i z)\right\}^{-1}[\hat{V}_i(0,p)] \qquad [7.22]$$

Now, the *modal transfer impedance matrix* $[\hat{Z}_{ti}(p)]$ seen from the source is defined by

$$[\hat{V}_i(0,p)] = [\hat{Z}_{ti}(p)][\hat{J}_i(0,p)] \qquad [7.23]$$

Relations [7.21] give the following:

$$[\hat{Z}_{ti}(p)] = \left\{[1]ch(\gamma_i z) + [\hat{Z}(p)][\hat{Z}_{ci}]^{-1}sh(\gamma_i z)\right\}^{-1}\left\{[\hat{Z}(p)]ch(\gamma_i z) + [\hat{Z}_{ci}]sh(\gamma_i z)\right\} \quad [7.24]$$

7.1.2.2. Impedance matching corresponding to a given Foucault modal current

A *matching of impedances* is defined by the condition

$$[\hat{Z}(p)] = [\hat{Z}_{ci}] \quad \text{then} \quad [\hat{Z}_{ti}] = [\hat{Z}_{ci}]$$

which is necessarily tied to a single mode i. Since the matrix $[Z_{ci}]$ contains elements made up of square roots, there is no network consisting of a finite number of lumped parameter components. Nevertheless, a network can be found having an infinite number of such elements.

Let us start with the case of a single line. We can write the scalar equation $\hat{Z}_{ci}(p) = \zeta_{ci}(p)Z_{c0}$, where $Z_{c0} = 1/vC$, while relation [7.14] giving ζ_{ci} is always valid. We obtain the above impedance by means of an infinite set of resistors and capacitors that are designed by means of an expansion of $\zeta_{ci}(p)$ into an infinitely continued fraction.

The following is the method to use. Let us rewrite [7.14] as follows:

$$\zeta_{ci}(p) = \zeta_{si}\sqrt{\frac{p+a_i}{p+b_i}}$$

where

$$\zeta_{si} = \sqrt{\mu_0 \varepsilon_d - \frac{\varepsilon_c^2 \beta_i^2}{\sigma_c^2}}, \quad a_i = \frac{\dfrac{\varepsilon_c \beta_i^2}{\sigma_c}}{\mu_0 \varepsilon_d - \dfrac{\varepsilon_c^2 \beta_i^2}{\sigma_c^2}}, \quad \text{and} \quad b_i = \frac{\sigma_d}{\varepsilon_d}$$

We have to use the following algebraic transformations:

$$[\zeta_{si}(p)]^2 = \zeta_{si}^2 + \underbrace{\frac{1}{\dfrac{b_i}{(a_i - b_i)\zeta_{si}^2} + \dfrac{p}{(a_i - b_i)\zeta_{si}^2}}}_{K_i(p)}$$

then

$$\zeta_{si}(p) - \zeta_{si} = \frac{1}{K_i(p)[\zeta_{si}(p) + \zeta_{si}]} = \frac{1}{2K_i(p)\zeta_{si} + \dfrac{1}{\zeta_{si}(p) + \zeta_{si}}}$$

so a recursive relation can be built easily by means of writing

$$\zeta_{si}(p) + \zeta_{si} = 2\zeta_{si} + \frac{1}{2K_i(p)\zeta_{si} + \dfrac{1}{\zeta_{ci}(p) + \zeta_{si}}}$$

Let us note here that $2K_i(p)\zeta_{si}$ is homogenous to the inverse of an impedance denoted as $\zeta_{mi}(p)$:

$$\zeta_{mi}(p) = \frac{1}{2K_i(p)\zeta_{si}}$$

To conclude, we obtain the infinitely continued fraction

$$\zeta_{ci}(p) = \zeta_{si} + \cfrac{1}{\cfrac{1}{\zeta_{mi}(p)} + \cfrac{1}{2\zeta_{si} + \cfrac{1}{\cfrac{1}{\zeta_{mi}(p)} + \cfrac{1}{2\zeta_{si} + \cdots}}}}$$

Going from $\zeta_{ci}(p)$ to $\hat{Z}_{ci}(p)$ is done by changing ζ_{si} into $Z_{si} = \zeta_{si}/vC$, then $K_i(p)$ into $K_i v^2 C^2$, and $\zeta_{mi}(p)$ into $\zeta_{mi}(p)/vC$. The required electrical scheme so designed is shown in Figure 7.1.

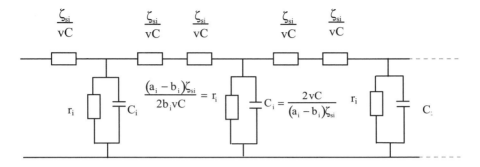

Figure 7.1. *Electrical scheme of a lossy impedance characteristic*

In the general case of coupled lossy lines, the capacitance C has to be replaced progressively by each coupling capacitance between each couple of lines involved in the whole network of transmission lines.

In the case of infinitely long lines, we obtain $[\hat{Z}_{ti}] = [\hat{Z}_{ci}]$. Then

$$[\hat{J}_i(0, p)] = [\hat{Z}_{ci}]^{-1}[\hat{V}_i(0, p)]$$

and

$$[\hat{V}_i(z, p)] = [\text{ch}(\gamma_i z) - \text{sh}(\gamma_i z)][\hat{V}_i(0, p)] = e^{-\gamma_i z}[\hat{V}_i(0, p)] \qquad [7.25]$$

which is a *forward progressive dispersive wave*.

7.1.2.3. *Scattering matrix for each Foucault modal current*

Now, in the general case, we can obtain the sharing in progressive and standing waves directly from relation [7.22] as follows:

$$[\hat{V}_i(0, p)] = \underbrace{[\hat{Z}_{ci}][\hat{Z}(p)]^{-1} e^{\gamma_i z}[\hat{V}_i(z, p)]}_{\text{progressive wave}} + \underbrace{\left\{[1] - [\hat{Z}_{ci}][\hat{Z}(p)]^{-1}\right\} \text{ch}(\gamma_i z)[\hat{V}_i(z, p)]}_{\text{standing wave}}$$

$$[7.26]$$

Let us finish by obtaining the expansion of any wave in progressive waves from relation [7.21] written as

$$[\hat{V}_i(z,p)] = 4[S_i]\{[1]+[S_i]\}^{-1}\{[1]+[S_i]e^{-2\gamma_i z}\}^{-1}[\hat{V}_i(0,p)]e^{-\gamma_i z}$$

where the matrix $[S_i] = \{[Z(p)]+[Z_{ci}]\}^{-1}\{[Z(p)]-[Z_{ci}]\}$ is the scattering matrix of the lossy coupled lines having the same length.

Going further with the previous expansion up to an infinite series as follows:

$$[\hat{V}_i(z,p)] = 4[S_i]\{[1]+[S_i]\}^{-1}\left\{\sum_{k=0}^{\infty}(-1)^k[S_i]^k e^{-2k\gamma_i z}\right\}[\hat{V}_i(0,p)]e^{-\gamma_i z}$$

leads to the expression of the wave having undergone n reflections:

$$[\hat{V}_i^{(n)}(z,p)] = 4(-1)^n[S_i]\{[1]+[S_i]\}^{-1}[S_i]^n[\hat{V}_i(0,p)]e^{-(2n+1)\gamma_i z}$$

7.2. Coming back in the time domain

7.2.1. *Inverse Laplace transform for lossy transmission lines*

For applications, it is necessary to come back in the time domain. This is achieved by means of the inverse Laplace transform. The lines of the vector rows $[\hat{V}(z,p)]$ and $[\hat{J}(z,p)]$ are "analytical" functions of the complex variable p almost everywhere in the complex plane outside some stand-alone points which are set on the left side of a straight line parallel to the imaginary axis. In this case, the inverse Laplace transform is applied as follows:

$$[7.4] \Rightarrow \begin{bmatrix}V(z,t)\\J(z,t)\end{bmatrix} = \frac{1}{2\pi j}\oint\begin{bmatrix}\hat{V}(z,p)\\\hat{J}(z,p)\end{bmatrix}e^{pt}\,dp \quad [7.27]$$

The integration path is a closed curve without any double point (this does not cut itself) in the complex plane of the variable p covered in the reverse clockwise direction and formed around all the singular points of the analytical functions involved in the second part of [7.27]. The analytical functions involved have to meet two constraints:

– their moduli converge towards zero when the modulus of p increases indefinitely;

– they are uniform, meaning that they have only one complex value corresponding to a complex value of p.

The first constraint is easily met in the transmission lines where any traveling wave is made up of progressive waves the Laplace transforms of which have an asymptotic behavior like $\exp[-p(t-z/v)]$ when the modulus of p increases indefinitely. Since the singular points of these Laplace transforms lie in the left-hand half-plane where the real part of p is negative, the modulus of the Laplace transform converges towards zero when the modulus of p increases indefinitely if and only if $z \leq vt$. This corresponds to the known causality tied to the finite velocity of light in the theory of relativity.

In contrast to this, the second constraint is never easily met with the application to distributed parameter networks like transmission lines. Indeed, progressive waves have been obtained in any wave traveling along a transmission line. These progressive waves show expressions containing the functions $e^{\pm \gamma z}$ where the complex parameter of propagation γ has the form $(1/v)\sqrt{(p+a)(p+b)}$, a and $b \geq 0$, while the standing waves, being even functions of γz, do not contain any square root in their mathematical expressions.

Let us point out the case of lossy progressive waves where the numbers a and b are different so that the complex parameter of propagation contains effectively a square root which leads to many-valued functions. Indeed, the complex number $(p+a)(p+b)$ defines a single-valued function of p because it has both a single modulus and argument set between 0 and 2π.

In contrast, its square root has two complex values having the same modulus but different arguments between 0 and 2π, which gives a two-valued function which does not reach the same value when the point p forms a closed curve without any double point.

7.2.2. Method of the contribution of loops

The way of getting a single-valued function is to draw a closed curve having *loops* made of two straight parallel and linked segments called *lips*, as illustrated in Figure 7.2.

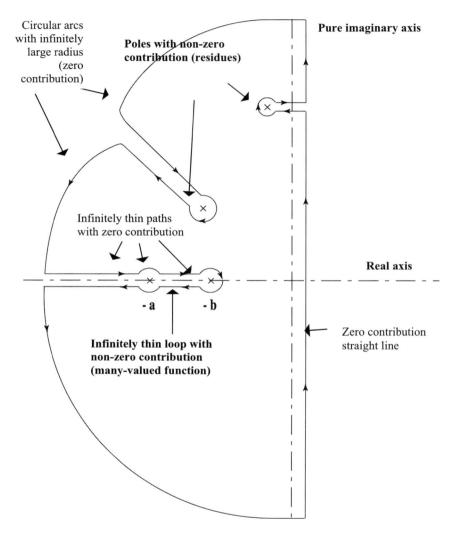

Figure 7.2. *Integration path of the inverse Laplace transform*

Then, the inverse Laplace transform gives the sum of two contributions:

signal waveform = pole residues + integrals along loops

the first being related to the *steady state* while the second is always related to a *transient state*.

How do we proceed with the computation of the time-domain waveform?

The contribution of poles is based on the very classical method of residues and its calculation is not difficult, being widely documented in the literature.

The contribution of loops, which is the difficult step in transmission line analysis, is pointed out now according to the behavior of γ along the lips of the loop. Let $a > b > 0$ for instance and set $p' = p + b$. On the upper lip of the loop very close to the real axis, we have

$$p \text{ real} \leq -b \quad \text{then } p' \text{ real} \leq 0 \quad \text{with } p \text{ real} \geq -a$$

Setting $p' = -x$, $x > 0$, we obtain $p + a = a - b - x \geq 0$. Thus, $(p+a)(p+b) = x(a-b-x)e^{j\pi}$ and

$$\gamma = \frac{1}{v}\sqrt{x(a-b-x)}e^{j\pi/2} \qquad [7.28]$$

Going from the upper lip to the lower lip is done by a rotation having the angle (-2π) around the point $(-b)$, which gives $p' = xe^{-2\pi j} = xe^{-\pi j}$ with $x > 0$ and $a - b - x > 0$, from which we obtain

$$\gamma = \frac{1}{v}\sqrt{x(a-b-x)}e^{-j\pi/2} \qquad [7.29]$$

Therefore, the value of γ has a jump when going through the real axis along the loop, which explains the non-zero contribution of the loop in the inverse Laplace transform.

What happens outside the loop?

Here, $(p + a)(p + b)$ real > 0. At the left side of the loop: $p + a = a - b - x < 0$, $p + b < 0$, so that we can write $(p+a)(p+b) = x(a-b-x)e^{2j\pi} > 0$; thus

$$\gamma = \frac{1}{v}\sqrt{x(a-b-x)}e^{j\pi}$$

above the real axis and

$$\gamma = \frac{1}{v}\sqrt{x(a-b-x)}e^{-j\pi}$$

below it; the contribution at the left side is zero.

Otherwise, the contributions are also zero, as shown in Figure 7.2.

In summary, the lossy infinitely long lines covered by progressive waves have transient states related to contributions of loops in the integration paths of the inverse Laplace transform in contrast to all the other lines covered by standing waves. This is illustrated below.

7.2.3. *Application to the distortion of a Dirac pulse in lossy media*

Let us begin with a *forward progressive wave* traveling on a lossy line. The transient state of the waveform can be obtained by means of feeding the line with a Dirac pulse. Then, it is well known that the transient state tied to any input signal is the result of the *convolution* product of it with the Dirac transient state of the line. This is easily known in the Laplace domain because a Dirac pulse at the line input has the image $\hat{V}(0, p) = 1$ and we get at the line output the image $e^{-\gamma_i z}$ of which we would like to know the original function in the time domain. This is the scope of the following discussion.

Applying the method of the integration loop developed previously, we are led to the following integral where it is assumed $a > b$ and which is valid only if $z \leq vt$:

$$\frac{1}{2\pi j} \oint_{\text{loop figure 12}} e^{pt-\gamma z} \, d\rho = \frac{1}{2\pi} \int_0^{a-b} e^{-(b+x)t} \sin\left[\frac{z}{v}\sqrt{x(a-b-x)}\right] dx \qquad [7.30]$$

As $x(a-b-x) \leq (a-b)/2$ when $0 \leq x \leq a-b$, this integral is less than $(e^{-t} - e^{-at})/2\pi t$, which is a strongly decreasing function showing a transient state as expected. Its duration t can be estimated by solving recursively the equation $e^{-t} - e^{-at} = \varepsilon t$, ε being a given number as small as possible, a and b are, respectively, the greatest and smallest numbers σ/ε and rv^2/c_i according to [7.7], or

$$\frac{\sigma_d}{\varepsilon_d} \quad \text{and} \quad \beta_i^2 \frac{\varepsilon_c}{\sigma_{ci}} \left(\mu_0 \varepsilon_d - \beta_i^2 \frac{\varepsilon_c^2}{\sigma_{ci}^2} \right)^{-1}$$

according to [7.11].

7.2.4. Classical kernel of the convolution methods

It is interesting to compute the above integral by means of known special functions.

Let us change the variable of integration:

$$x = \frac{a-b}{2}(1+\cos\theta)$$

This yields

$$-\frac{a-b}{2\pi} e^{-\frac{a+b}{2}t} \int_0^\pi e^{-\frac{a-b}{2}t\cos\theta} \sin\left(\frac{a-b}{2}\frac{z}{v}\sin\theta\right) \sin\theta\, d\theta$$

Writing

$$t\cos\theta + j\frac{z}{v}\sin\theta = \sqrt{t^2 - \frac{z^2}{v^2}}\cos(\theta - \theta_0) \quad \text{with} \quad \tan\theta_0 = -j\frac{z}{vt}$$

we are led to

$$[F(\theta_0) - F(-\theta_0)]\frac{a-b}{4\pi j} e^{-[(a+b)/2]t}$$

with the function F being defined by

$$F(\theta_0) = \int_{-\theta_0}^{-\theta_0+\pi} \exp\left(-\frac{a-b}{2}\sqrt{t^2 - \frac{z^2}{v^2}}\cos\varphi\right)(\cos\varphi\sin\theta_0 + \sin\varphi\cos\theta_0)\, d\varphi$$

Let

$$u = \frac{b-a}{2}\sqrt{t^2 - \frac{z^2}{v^2}}$$

and using the classical formula of the Bessel modified functions:

$$e^{u\cos\varphi} = I_0(u) + 2I_1(u)\cos\varphi + 2I_2(u)\cos 2\varphi + \cdots$$

gives the expansion

$$F(\theta_0) = 2I_0(u)(2\cos\theta_0 \sin\theta_0 + 2\sin\theta_0 \cos\theta_0) + \frac{\pi}{2}I_1(u)\sin\theta_0$$

$$+2I_2(u)\left[\cos\theta_0\left(\frac{\cos 3\theta_0}{3} - \cos\theta_0\right) + \sin\theta_0\left(\frac{\sin 3\theta_0}{3} - \sin\theta_0\right)\right] + \cdots$$

So, we obtain $F(\theta_0) - F(-\theta_0) = \pi \sin\theta_0 I_1(u)$, the other terms being zero.

Thus, the final result is

$$\frac{a-b}{4}\frac{z}{v}e^{-[(a+b)/2]t}\frac{I_1\left(\frac{a-b}{2}\sqrt{t^2 - \frac{z^2}{v^2}}\right)}{\sqrt{t^2 - \frac{z^2}{v^2}}} = N(z,t)$$

which represents the kernel of the convolution methods.

Then, any output signal is obtained by means of the *convolution product* $s(z,t) = N(z,t) * e(0,t)$, or more explicitly:

$$s(z,t) = \int_0^t N(z,t')e(0, t-t')\,dt' \qquad [7.31]$$

7.2.5. *Diffusion equation and the time-varying "skin depth"*

What is the asymptotic behavior of this function? It is necessary to use the asymptotic formula of the Bessel functions:

$$I_1(u) = \frac{e^u}{\sqrt{2\pi u}}\left[1 + o\left(\frac{1}{u}\right)\right]$$

so that

$$N(z,t) = \frac{a}{4}\frac{z}{v}e^{-\frac{a}{2}t}\frac{I_1\left(\frac{a}{2}\sqrt{t^2-\frac{z^2}{v^2}}\right)}{\sqrt{t^2-\frac{z^2}{v^2}}} = \frac{a}{4}\frac{z}{v}e^{-\frac{a}{2}t}\frac{e^{\frac{a}{2}\sqrt{t^2-\frac{z^2}{v^2}}}}{\sqrt{2\pi}\left(t^2-\frac{z^2}{v^2}\right)^{\frac{3}{4}}}\left[1+o\left(\frac{1}{t}\right)\right]$$

The asymptotic expansion

$$\sqrt{t^2-\frac{z^2}{v^2}} = t\left(1-\frac{z^2}{2v^2t^2}+o\left(\frac{1}{t}\right)\right)$$

when $t \to \infty$, leads to

$$N(z,t) = \frac{a}{4}\frac{z}{vt^{3/2}}e^{-az^2/4tv^2}\left[1+o\left(\frac{1}{t}\right)\right]$$

This corresponds classically to the solution of the *diffusion equation*:

$$D\frac{\partial^2 f}{\partial z^2} - \frac{\partial f}{\partial t} = 0$$

where the diffusion parameter D is equal to v^2/a [LAN3]. The diffusion equation shows a wave edge running according to the law $z = 2v\sqrt{t/a}$: this is the *time-varying "skin depth"*.

7.2.6. Multiple reflections processing

In the general case of any line having a finite length, there are multiple reflections and the evolution of a Dirac pulse is obtained from

$$V_i(z,t) = \frac{1}{2\pi j}\oint_{\text{loop figure 12}} e^{pt}\frac{dp}{\text{ch}(\gamma_i z) + \frac{\hat{Z}_{ci}}{Z(p)}\text{sh}(\gamma_i z)}$$

Here, the expression behind the integral is an even function of $\gamma_i z$ and, therefore, no loop is needed in the integration path. Only contributions of poles are of concern, these poles being solutions of the equation

$$\text{ch}\left[\frac{z}{v}\sqrt{(p+a)(p+b)}\right]+\frac{\hat{Z}_{ci}(p)}{Z(p)}\text{sh}\left[\frac{z}{v}\sqrt{(p+a)(p+b)}\right]=0$$

The solution p_k being a pole having the residue r_k, the Cauchy theorem of analytical function residues leads directly to the result

$$N(z,t)=\sum_k r_k e^{p_k t} \qquad [7.32]$$

which is related to the standing waves.

This is easily generalized to any set of lossy coupled lines having the same length:

$$[V_i(z,t)]=\frac{1}{2\pi j}\oint_{\text{loop figure 12}} e^{pt}\left\{[1]\text{ch}(\gamma_i z)+[\hat{Z}_{ci}][\hat{Z}(p)]^{-1}\text{sh}(\gamma_i z)\right\}^{-1}[\hat{V}_i(0,p)]\,dp$$

The poles p_k are solutions of the equation

$$\det\left\{[1]\text{ch}(\gamma_i z)+[\hat{Z}_{ci}][\hat{Z}(p)]^{-1}\text{sh}(\gamma_i z)\right\}=0$$

and the final result is

$$[N(z,t)]=\sum_k [r_k]e^{p_k t}$$

Nevertheless, in spite of the case of pure resistive loads, pole computations cannot be done analytically; then it is necessary to deal with the discrete Fourier transform that we discuss now.

7.3. Method of the discrete Fourier transform

7.3.1. *Fourier transform and the harmonic steady state*

For physically useful signals having either a finite energy or periodic distributions, the Fourier transform is classically involved. Then, the input sources can be represented as

$$\begin{bmatrix} [V_i(0,t)] \\ [J_i(0,t)] \end{bmatrix} = \frac{1}{2\pi} \int_{-\infty}^{+\infty} e^{j\omega t} \begin{bmatrix} [FV_i(0,\omega)] \\ [FJ_i(0,\omega)] \end{bmatrix} d\omega \quad [7.33]$$

where F defines the operator of the Fourier transform which is the mapping from the time domain to the frequency domain of the variable ω.

Recalling that the image of the function $e^{j\omega t}$ in the Laplace domain is $1/(p-j\omega)$, we obtain

$$\begin{bmatrix} [\hat{V}_i(0,p)] \\ [\hat{J}_i(0,p)] \end{bmatrix} = \int_0^{+\infty} e^{-pt} \begin{bmatrix} [V_i(0,t)] \\ [J_i(0,t)] \end{bmatrix} dt = \int_{-\infty}^{+\infty} \begin{bmatrix} [FV_i(0,\omega)] \\ [FJ_i(0,\omega)] \end{bmatrix} \frac{d\omega}{p-j\omega} \quad [7.34]$$

Applying transfer relation [7.19] then inverse Laplace transform [7.27], we obtain

$$\begin{bmatrix} [V_i(0,t)] \\ [J_i(0,t)] \end{bmatrix} = \frac{1}{2\pi j} \oint e^{pt} [T_i(z,p)] \begin{bmatrix} [V_i(0,p)] \\ [J_i(0,p)] \end{bmatrix} dp$$

After exchanging the integrations, we are led to compute first the integration with respect to the variable p:

$$\frac{1}{2\pi j} \oint e^{pt} [T_i(z,p)] \frac{dp}{p-j\omega} = [T_i(z,t,\omega)] \quad [7.35]$$

and then that with respect to the frequency ω:

$$\begin{bmatrix} [V_i(z,t)] \\ [J_i(z,t)] \end{bmatrix} = \frac{1}{2\pi} \int_{-\infty}^{+\infty} [T_i(z,t,\omega)] \begin{bmatrix} [\bar{F}V_i(0,\omega)] \\ [\bar{F}J_i(0,\omega)] \end{bmatrix} d\omega \quad [7.36]$$

Considering the first integration [7.35]: there is only the contribution of the pole $p = j\omega$. Its residue is equal to $[T_i(z,j\omega)]e^{j\omega t}$ that defines the *harmonic steady state*. So, integral [7.36] corresponds to the classical *Fourier transform*.

7.3.2. Discrete Fourier transform and the sampling procedure

Practically, two interesting cases can be highlighted. The first concerns *harmonic signals*. Their spectrum is made up of discrete frequencies $\omega = \omega_k$ which are Dirac distributions as follows:

$$FV_i(0,\omega) = \sum_k a_k \delta(\omega - \omega_k)$$

which gives the time-domain signal $V_i(0,t) = \sum_k a_k e^{j\omega_k t}$ for each discrete frequency $\omega = \omega_k$. Then, we obtain the transfer matrix $[Ti(z, j\omega_k)]$, corresponding to each discrete frequency, inside which the complex parameter of propagation $\gamma(j\omega_k)$ occurs as a complex number $\alpha(\omega_k) + j\beta(\omega_k)$, where α defines the *losses parameter* per unit length for both the dielectrics and conductors and β the *phase difference* per unit length.

The *phase velocity* is defined by means of

$$v_\varphi = \frac{\omega_k}{\beta(\omega_k)}$$

while the *group velocity* is equal to

$$v_g = \left(\frac{\partial \omega}{\partial \beta}\right)_{\omega = \omega_k}$$

The other case concerns the signals having a finite duration T. Their spectrum is given by the *Shannon interpolation formula*:

$$FV_i(0,\omega) = \sum_{k=-\infty}^{k=+\infty} FV_i\left(0, \frac{2\pi k}{T}\right) \frac{\sin\left(\omega \frac{T}{2} - k\pi\right)}{\omega \frac{T}{2} - k\pi}$$

according to which the spectrum is completely defined by knowing its values at the discrete frequencies $\omega_k = 2k\pi/T$. Coming back to the time domain is achieved by the discrete Fourier transform:

$$V_i(0,t) = \frac{2}{T} \sum_{k=-\infty}^{k=+\infty} FV_i\left(0, \frac{2\pi k}{T}\right) e^{-2\pi jk(t/T)} \qquad [7.37]$$

Then, we obtain the so-called *sampling procedure* of the signal.

This can be numerically estimated by means of the truncated expansion denoted as $(PV_i)(0, t)$:

$$(PV_i)(0,t) = \frac{2}{T} \sum_{k=-N}^{k=+N} FV_i\left(0, \frac{2\pi k}{T}\right) e^{-2\pi jk(t/T)} \qquad [7.38]$$

7.3.3. *Application to digital signal processing*

We need to estimate N for the digital signals shown in Figure 7.3.

Recalling the classical Fourier transform of a triangular signal being zero outside a given segment of length 2τ and reaching the value 1 at its top point:

$$F\left(\underset{-\tau \ \ +\tau}{\wedge}\right) = \frac{\tau}{4}\left(\frac{\sin\frac{\omega\tau}{2}}{\frac{\omega\tau}{2}}\right)^2$$

the Fourier transform of the digital signal is obtained easily:

$$FV_i(\omega) = \sum_{k=-M}^{k=+M} e^{-jk\omega\tau} \frac{\tau}{4}\left(\frac{\sin\frac{\omega\tau}{2}}{\frac{\omega\tau}{2}}\right)^2 = 4\frac{\sin\frac{\omega T}{2}\sin\frac{\omega\tau}{2}}{\tau\omega^2}$$

The inverse of τ represents the "slew rate" of the linear signal edge ds/dt.

Let us estimate the spectrum range $|\omega| \leq \Omega_0$ of the digital signal by means of having a given ratio of its energy set in the spectrum range. Applying the classical *Parseval theorem*, we obtain

$$\int_{-\infty}^{+\infty}[V_i(t)]^2 \, dt = \frac{1}{2\pi}\int_{-\infty}^{+\infty}[FV_i[\omega]]^2 \, d\omega$$

214   Electromagnetism and Interconnections

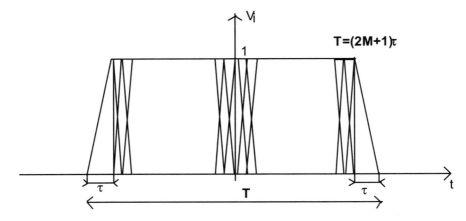

Figure 7.3. *Digital signal in the time domain*

Thus, we have to define the spectrum range $|\omega| \leq \Omega_0$ so that

$$\int_{|\omega|>\Omega_0} [FV_i[\omega]]^2 \, d\omega \leq \varepsilon \int_{-\infty}^{+\infty} [FV_i[\omega]]^2 \, d\omega = 2\pi\varepsilon \int_{-\infty}^{+\infty} [V_i[t]]^2 \, dt$$

We easily compute

$$\int_{-\infty}^{+\infty} [V_i[t]]^2 \, dt = T - \frac{4\tau}{3} < T$$

while we can find merely an upper bound of the integral outside the spectrum range:

$$\int_{|\omega|>\Omega_0} [FV_i[\omega]]^2 \, d\omega = 32 \int_{\Omega_0}^{\infty} \frac{\sin^2 \frac{\omega T}{2} \sin^2 \frac{\omega \tau}{2}}{\tau^2 \omega^4} \, d\omega < \frac{32}{\tau^2} \int_{\Omega_0}^{\infty} \frac{d\omega}{\omega^4} = \frac{32}{3\tau^2 \Omega_0^3}$$

So, we are led to choose the constraint

$$\frac{32}{3\tau^2 \Omega_0^3} = 2\pi\varepsilon T$$

With $1/\tau = ds/dt$ and $\Omega_0 = 2\pi N/T$, which is the spectrum range, we obtain

$$N = 0.19\varepsilon^{-1/3}T^{2/3}\left(\frac{ds}{dt}\right)^{2/3} \qquad [7.39]$$

which defines the *truncature criterion* and

$$\frac{N}{T} = 0.19\varepsilon^{-1/3}T^{-1/3}\left(\frac{ds}{dt}\right)^{2/3}$$

which defines the *spectrum range*.

We meet the well-known property that the spectrum range is as wide as the "slew rate" is high and the duration of the signal is short.

7.3.4. *Bifurcations and complex networks of lossy transmission lines*

Let us return to Figure 5.11 concerning heterogenously coupled lines. Equation [5.24] has to be directly translated into the Laplace domain in order to be extended to lossy lines. In so doing, it becomes

$$\begin{bmatrix}[V'_i(z_1+z'_2,p)]\\ [J'_i(z_1+z'_2,p)]\\ [V''_i(z_1+z''_2,p)]\\ [J''_i(z_1+z''_2,p)]\end{bmatrix} - \begin{bmatrix}[T'_i](1-e^{-2\gamma_i z'_2}) & [0] \\ [0] & [T''_i](1-e^{-2\gamma_i z''_2})\end{bmatrix}\begin{bmatrix}[V'_i(z_1,p)]\\ [J'_i(z_1,p)]\\ [V''_i(z_1,p)]\\ [J''_i(z_1,p)]\end{bmatrix} = \begin{bmatrix}[0]\\ [0]\\ [0]\\ [0]\end{bmatrix} \qquad [7.40]$$

Also, the nodal operational matrices of Figure 6.2 have to be written directly in the Laplace domain by means of replacing the operators $\partial/\partial t$ by p, nevertheless with the constraint of zero initial conditions. Thus, equation [6.1] can be easily translated into the Laplace domain as above.

In the same way, bifurcation equations [5.25] become

$$\left.\begin{array}{l}[V_i^{(k)}(z,p)]-[V_i^{(1)}(z,p)]=[0]\\ \sum_k[J_i^{(k)}(z,p)]=[0]\end{array}\right\} \qquad [7.41]$$

Chaining all the matrix equations into a single one, so defining the vectors $\hat{X}$ and $\hat{Y}$ depending on $[z]$ in the Laplace domain, we are led to the general *state equation* corresponding to [5.26]:

$$\left[\hat{A}([z],p)\right]\left[\hat{X}([z],p)\right]=\left[\hat{Y}([z],p)\right] \qquad [7.42]$$

where $[\hat{A}([z],p)]$ is a matrix of which the elements are single-valued analytical functions without poles. Letting

$$[Y([z_0],t)] = \frac{1}{2\pi}\int_{-\infty}^{+\infty} e^{j\omega t} FY([z_0],\omega)\,d\omega$$

we obtain directly

$$[X([z],t)] = \int_{-\infty}^{+\infty} e^{j\omega t}\left[\hat{A}([z],\omega)\right]^{-1}\left[\hat{Y}_0([z],j\omega)\right]d\omega + \sum \text{pole residues of }\left[\hat{A}([z],p)\right]^{-1}$$

$$[7.43]$$

The poles are the solution of

$$\det\left[A([z],p)\right]=0 \qquad [7.44]$$

At this step of the computation, the discrete Fourier transform has to be applied for estimating the Fourier integral of the right-hand side of equation [7.43] where p is equal to $p_k = 2\pi j(k/T)$, $|k| \leq N$, T is the duration of the signal.

Application leads to the following:

– estimation of N defining the spectrum range according to [7.39];

– computation of the $(2N+1)$ complex vectors

$$\left[\hat{Y}\left([z_0], 2\pi j\frac{k}{T}\right)\right],\quad |k|\leq N$$

by means of the Gauss–Seidel algorithm processed $(2N+1)$ times in the equations

$$\left[\hat{A}\left([z], 2\pi j\frac{k}{T}\right)\right]\left[\hat{X}\left([z], 2\pi j\frac{k}{T}\right)\right]=\left[\hat{Y}\left([z_0], 2\pi j\frac{k}{T}\right)\right]\quad |k|\leq N \qquad [7.45]$$

– coming back to the time domain $[X([z],t)]$ according to the discrete Fourier transform [7.37] approximated by [7.38].

Nevertheless, it is necessary to take account of the fact that the poles do not obviously correspond to the discrete frequencies of the sampling procedure. So, in general the multiple reflections cannot be accounted for without tightening the discrete frequency steps, which leads to higher N than previously computed.

Let N be the number of segments used inside the network; the state vectors need $8L$ lines of real numbers. Therefore, applying the Gauss algorithm, the computation requirement is estimated to be equal to $(2N+1)(8L)^3/3$ elementary operations and the storage capacity is equal to $(2N+1)[(8L)^2 + 8L]$ real numbers in the case of forming curves or $2[(8L)^2 + 8L]$ real numbers in the case of getting the values of the signal at a given time.

Appendix D shows a modified Gauss–Seidel algorithm which is a recursive method of solving [7.45] by saving a large amount of storage capacity.

7.4. Conclusion

It has been shown how a matrix state equation in the frequency domain can be obtained for any homogenous interconnection media having both lossy dielectrics and lossy coupled conductors accounting for the Foucault modal currents. Nevertheless the spectrum range required by high-speed digital signals becomes wider and wider as their transition times become shorter and shorter which means longer and longer computation time for coming back in the time domain. Moreover, the progressive waveforms have to be computed by means of the method of the contribution of loops that leads to either integral or convolution products requiring large memory storage space.

Chapter 8

Time-domain Wavelets

Wavelets are presented in the very general framework of the theory of linear operators, particularly within *commutative operators algebra*, of which we use the strictly necessary elements suited for applications to lossy wave propagation with the aim of achieving numerical computations. Then, an efficient expansion of any signal transmitted along waveguides into wavelets accounting for boundary conditions is demonstrated. This is applied directly to lossy lines. A useful detailed development of the speed of convergence of the involved expansions is done for the high-speed digital signals needed for applications in electronics.

8.1. Theoretical introduction

8.1.1. *Motivation for the time-domain wavelets method*

Modeling signal propagation requires time-domain computation in order to account for the nonlinear behavior of electronic devices located at transmission line terminals.

Moreover, high-speed digital signals have a very wide spectrum range that leads to many computations in the frequency domain which is necessarily replaced by a convolution method requiring a huge data storage memory for nonlinear modeling.

These difficulties lead to the introduction of the time-domain "wavelets" solution of the lossy propagation equation.

These wavelets are different from those defined by Morlet and Meyer in signal processing [MEY].

Before presenting the application of wavelets to solving the lossy propagation equation, let us explain the mathematical framework inside which these wavelets can be defined by means of functional analysis, the strictly useful elements of which are briefly recalled here.

8.1.2. *General mathematical framework*

Let E be a vector space the elements of which are matrix functions defined in $\Omega \subset C^m$, C being the set of complex numbers, or matrix operators defined in A^n, A being an operators algebra, and having values in $C^m \times C^m$ or $A^m \times A^m$.

Let A be an operator, which is linear and compact, working in (E), which means it transforms any matrix function or operator belonging to (E) in an other one into (E).

Let $[A]$ be a matrix belonging to $C^m \times C^m$ or $A^m \times A^m$; we are led to solve the following equation in (E):

$$[A][u] = [\Lambda][u] \qquad [8.1]$$

where $[u]$ is a matrix function or operator belonging to (E).

The method of solving this equation lies in using commutative operators algebra [DAU4] which is a "Banach algebra" in so far as an involution and a norm of uniform convergence are defined and aimed at the numerical approximations of the required solution as discussed further later on.

Indeed, it is known, according to functional analysis, that the commutative linear operators belong to a set which has the properties of both vector spaces and of polynomial rings. Thus, inside these polynomial rings, the computations are done for the commutative (matrix) operators similar to those concerning numbers. Before going further, we make the simplification of omitting the brackets for ease of presentation, so $[u]$, $[A]$ and $[\Lambda]$ become u, A and α^2, respectively. Then, we have to solve

$$Au = \alpha^2 u \qquad [8.2]$$

by setting $u = B\Psi_0$, B being a linear operator belonging to the Banach space $\tilde{E}$ related to E, which can be expanded in series of powers of an operator denoted as a non-commutative product $\alpha^{-1}G$, where G is cleverly chosen according to the equation to be solved and does not depend on the particular physical conditions. We write the following expansion:

$$B = \sum_{n=0}^{\infty} a_n (\alpha^{-1}G)^n$$

where Ψ_0 is a solution of [8.2] belonging to (E) and not depending on the particular physical conditions. The operators $(\alpha^{-1}G)$ belong to the base of the vector space defined by the algebra $\tilde{E}$.

Entities a_n are mathematical features to be defined according to the required physical conditions to be satisfied by the solutions of equation [8.2]. This, then, is the technical target of the "wavelet" method in respect of the scope mentioned at the beginning of this section.

8.1.3. *Seed and generator of direct and reverse wavelets family*

8.1.3.1. *Direct wavelets*

Let us deal with the characterization of operator G versus operator A so that equation [8.2] can be solved efficiently in the most general case.

In order to do this, we set zero all the a_i of the B operator expansion except the one corresponding to $i = n$ assumed to be equal to 1. Then, we obtain $\Psi_n = (\alpha^{-1}G)^n \Psi_0$ which is a particular solution of [8.2]. So we write the following recursive relation which is valid for any $n \geq 1$:

$$\Psi_n = (\alpha^{-1}G)(\alpha^{-1}G)^{n-1}\Psi_0 = (\alpha^{-1}G)\Psi_{n-1} \qquad [8.3]$$

Note that multiplying on the left side by α yields $\alpha \Psi_n = G\Psi_{n-1}$.

Let G be such that $G\Psi_n$ becomes a solution of [8.3] whatever n is. We write

$$G\Psi_n = (\alpha^{-1}G)(G\Psi_{n-1}) = (\alpha^{-1}G)(\alpha\Psi_n)$$

However, multiplying the two parts of [8.2] on their left-hand side by G yields

$$G\Psi_n = G(\alpha^{-1}G)\Psi_{n-1} = (G\alpha^{-1})(G\Psi_{n-1}) = (G\alpha^{-1})(\alpha\Psi_n)$$

then $(G\alpha^{-1} - \alpha^{-1}G)(\alpha\Psi_n) = 0$, which means $(\alpha\Psi_n)$ belongs to the so-called kernel of the operator.

The operators $(\alpha^{-1}G)^n$ define a base of $\tilde{E}$ which is an algebra then a vector space, so Ψ_n are a base of the vector space E that leads to

$$(G\alpha^{-1} - \alpha^{-1}G)\alpha \sum_{n \geq 0} a_n \Psi_n = 0, \quad \forall a_n$$

which shows the kernel fills all of E itself. This is possible only if the operator having this kernel is zero, so

$$G\alpha^{-1} = \alpha^{-1}G \quad \text{then} \quad \alpha G = G\alpha$$

This is the first commutation relation.

Let us deal with the equation without any other assumption:

$$\alpha^2 \Psi_n = \alpha^2 (\alpha^{-1}G)\Psi_{n-1} = \alpha G \Psi_{n-1} = \alpha G(\alpha^{-2} A)\Psi_{n-1}$$

therefore $(\alpha G \alpha^{-2} A - A\alpha^{-1}G)\Psi_{n-1} = 0$, which means Ψ_{n-1} belongs to the so-called kernel of the operator $(\alpha G \alpha^{-2} A - A\alpha^{-1}G)$ which fills the whole space E as previously shown. This is possible only if

$$\alpha G \alpha^{-2} A = A\alpha^{-1}G$$

According to the first commutation relation:

$$\alpha G \alpha^{-2} A = \alpha \alpha^{-2} GA = \alpha^{-1} GA$$

Thus, we obtain the second *commutation relation*:

$$(\alpha^{-1}G)A = A(\alpha^{-1}G) \qquad [8.4]$$

Now we propose the following definition. A and α being two non-necessarily commutative linear operators working inside a (Banach) vector space (E), it is said that the solutions of the equation

$$Au = \alpha^2 u \qquad u \in E$$

are spanned by the *wavelets family* $\{\Psi_0, \Psi_1, \ldots \Psi_n, \ldots\}$ having the *seed* Ψ_0 and the *generator* G if and only if

$$\alpha \Psi_n = G \Psi_{n-1}, \quad n \geq 1, \quad n \in N$$

Then G and α commute, and $\alpha^{-1} G$ and A commute.

8.1.3.2. Reverse wavelets

This result leads naturally to the family of reverse wavelets denoted as Ψ'_n having the same seed as the direct wavelets and defined by the generator (non-commutative product):

$$G' = G^{-1} A \qquad [8.5]$$

according to the recursive relation

$$\Psi'_n = \alpha^{-1} G' \Psi'_{n-1} = (\alpha^{-1} G')^n \Psi_0 \qquad [8.6]$$

Then, we get the *non-commutative factorization*

$$A = GG' \qquad [8.7]$$

8.1.3.3. Wavelets tied to the commutative factorization of a linear operator

Let us show the actions of the direct generator G and the reverse generator G' on a direct wavelet Ψ_n and on a reverse wavelet Ψ'_n, respectively, in the important case of the commutability $GG' = G'G$.

This occurs in electromagnetism with the factorization of the operator ∇^2 into the product of the vector operator $\vec{\nabla} \vee$ for the vector function having a zero divergence.

Concerning Ψ_n, relation [8.3] yields

$$G \Psi_n = \alpha \Psi_{n+1} \quad \text{(increasing index)}$$

Moreover, applying the first commutation relation to the reverse generator G' leads to

$$GG' \Psi_n = A \Psi_n = \alpha^2 \Psi_n = G'G \Psi_n = G' \alpha \Psi_{n+1} = \alpha G' \Psi_{n+1}$$

so

$$\alpha(G'\Psi_{n+1} - \alpha\Psi_n) = 0$$

thus

$$G'\Psi_n = \alpha\Psi_{n-1} \quad \text{(decreasing index)} \qquad [8.8]$$

This relation allows us to write $\Psi_{-n} = \alpha^{-1}\Psi_{-n+1} = (\alpha^{-1}G)^n\Psi_0 = \Psi'_n$.

In the same way, if $G'\Psi'_n = \alpha\Psi'_{n+1}$, we get $G\Psi'_n = \alpha\Psi'_{n-1}$ and

$$\Psi'_n = (G^{-1}\alpha)\Psi'_{n-1} = (G^{-1}\alpha)^n\Psi_0 = (\alpha^{-1}G)^{-n}\Psi_0 = \Psi_{-n}$$

This shows the single seed Ψ_0 can generate in one way both the direct and reverse wavelets, the first one being defined by a positive index and the second one being defined by a negative index. So, the set of wavelets has a *group structure*.

By means of the *commutative factorization of the operator A* which then belongs to a "von Neumann algebra", a *family of "direct" wavelets* can be found attached to a *"direct" generator G* and a *family of "reverse" wavelets* attached to an *"inverse" generator G'*, both families having the same *seed* Ψ_0. The direct wavelets are related to *incident waves*, while the reverse wavelets are related to *reflected waves*.

The seed Ψ_0 solution of equation [8.2] can be computed by means of the Banach fixed point theorem as will be discussed later on in terms of the Picard method applied to guided wave propagation.

8.1.3.4. *Convergence and numerical approximation of a wavelets expansion*

Let us deal with the expansion of any solution into the wavelet base. Considering the direct wavelet expansion, we write

$$u = \sum_{n=0}^{\infty} a_n \Psi_n$$

This is useful if and only if this expansion is the only one; this is the case because of the property of any vector space base, on the one hand, and if it is convergent, which we point out now, on the other hand.

So, any wavelet family can be called a "Schauder base" of the space E.

For the sake of numerical computation, the last expansion has to be truncated at the level of its first n terms. Let P_n be the *projector* or *linear operator of projection* onto a subspace defined by

$$u \xrightarrow{P_n} P_u = \sum_{k=0}^{k=n-1} a_k \Psi_k$$

The computation error is given by $u - P_n u$. So, the criterion of stopping is based on using the norm of the subspace as follows:

$$\begin{aligned}
\|u - P_n u\| &= \left\|\sum_{k=n}^{\infty} a_k \Psi_k\right\| \\
&\leq \sum_{k=n}^{\infty} \|a_k \Psi_k\| \quad \text{(according to the triangle inequality)} \\
&\leq \sum_{k=n}^{\infty} |a_k| \|\Psi_k\| \quad \text{(property of norms)} \\
&\leq \sum_{k=n}^{\infty} |a_k| \|(\alpha^{-1}G)^n \Psi_{k-n}\| \quad \text{(because of [8.3])}
\end{aligned}$$

According to the definition of the norm of a linear operator, we obtain

$$\|(\alpha^{-1}G)^n\| = \sup_{\Psi \in E} \frac{\|(\alpha^{-1}G)^n \Psi\|}{\|\Psi\|}$$

with the following inequalities:

$$\|(\alpha^{-1}G)^n \Psi_{k-n}\| \leq \|(\alpha^{-1}G)^n\| \|\Psi_{k-n}\|$$
$$\|\Psi_{k-n}\| = \|(\alpha^{-1}G)^{k-n} \Psi_0\| \leq \|(\alpha^{-1}G)^{k-n}\| \|\Psi_0\|$$

Then the stopping criterion corresponds to the smallest integer n so that

$$\|u - P_n u\| \leq \|\Psi_0\| \|(\alpha^{-1}G)^n\| \sum_{m=0}^{\infty} |a_{m+n}| \|(\alpha^{-1}G)^m\| < \varepsilon \qquad [8.9]$$

where ε is a number as small as needed.

226 Electromagnetism and Interconnections

8.2. Application to digital signal propagation

8.2.1. *Application to lossless guided wave analysis in the time domain*

8.2.1.1. *Commutative factorization of the lossless guided waves operator*

Any equation of *lossless guided waves* along the z axis can be written as

$$\left(v^2 \frac{\partial^2}{\partial z^2} - \frac{\partial^2}{\partial t^2} + \alpha^2\right) u = 0 \qquad [8.10]$$

where α is a complex number. This equation corresponds to equations [4.7], [4.13], [4.25], and [4.31] with $\sigma = 0$, previously encountered in the modal analysis of electromagnetic fields where

$$\alpha = j\beta_i$$

In this case, the plane $\mathbf{R}^2$ is (z, t). Here, the operator A met in the previous section is as follows:

$$A = -v^2 \frac{\partial^2}{\partial z^2} + \frac{\partial^2}{\partial t^2} = GG'$$

with the direct generator

$$G = v\frac{\partial}{\partial z} + \frac{\partial}{\partial t}$$

and the reverse generator

$$G' = -v\frac{\partial}{\partial z} + \frac{\partial}{\partial t}$$

8.2.1.2. *Seed of a wavelet base*

Let us compute the direct wavelets of the family $\{\Phi_0, \Phi_1, \ldots \Phi_n, \ldots\}$ related to the generator G according to the relations $\Phi_n = (\alpha^{-1}G)\Phi_{n-1} = (\alpha^{-1}G)^n \Phi_0$ previously found.

We have to start by computing Φ_0 by means of the application of the *Banach fixed point theorem* to the partial differential equation [8.10] according to the *classical Picard method* mentioned above.

Let us return to the transformation of variables already used for solving the lossless propagation equation:

$$\begin{cases} \xi = \frac{1}{2}\left(t + \frac{z}{v}\right) \\ \xi' = \frac{1}{2}\left(t - \frac{z}{v}\right) \end{cases} \text{ gives } \begin{cases} G = \frac{\partial}{\partial \xi} \\ G' = \frac{\partial}{\partial \xi'} \end{cases}$$

and the equation becomes

$$\frac{\partial^2 u}{\partial \xi \partial \xi'} \alpha^2 u$$

Defining the wavelet Φ_0 with the condition $u(0, 0) = 1$, we can write the following integral equation arising from the last differential equation:

$$\Phi_0(\xi, \xi') = \alpha^2 \int_0^\xi \left[\int_0^{\xi'} \Phi_0(\lambda, \mu) \, d\mu \right] d\lambda + 1$$

The *Picard method* relies on the recursive relation

$$\Phi_0^{(n+1)}(\xi, \xi') = \alpha^2 \int_0^\xi \left[\int_0^{\xi'} \Phi_0^{(n)}(\lambda, \mu) \, d\mu \right] d\lambda + 1$$

with the starting point:

$$\Phi_0^{(0)}(\xi, \xi') = \Phi_0(0, 0) = 1$$

going on step by step:

$$\Phi_0^{(1)}(\xi, \xi') = \alpha^2(\xi, \xi') + 1$$
$$\vdots$$

up to:

$$\Phi_0^{(n)}(\xi,\xi') = \sum_{k=0}^{k=n} \alpha^{2k} \frac{(\xi\xi')^k}{(k!)^2} = \sum_{k=0}^{k=n} \frac{(2\alpha\sqrt{\xi\xi'})^{2k}}{(2^k k!)^2}$$

While n grows indefinitely, $\Phi_0^{(n)}(\xi,\xi')$ converges towards the solution:

$$\Phi_0(\xi,\xi') = I_0(2\alpha\sqrt{\xi\xi'}) = I_0\left(\alpha\sqrt{t^2 - \frac{z^2}{v^2}}\right) \qquad [8.11]$$

where I_0 is the zero-order *modified Bessel function* of the first kind.

8.2.1.3. *Direct wavelets*

Knowing Φ_0, all the other wavelets are directly computed as follows:

$$\Phi_n = \frac{1}{\alpha} \frac{\partial \Phi_{n-1}}{\partial \xi} = \alpha^{-n} \frac{\partial^n \Phi_0}{\partial \xi^n}, \quad n \geq 1$$

By means of

$$\Phi_0 = I_0(2\alpha\sqrt{\xi\xi'}) = \sum_{k=0}^{\infty} \alpha^{2k} \frac{(\xi\xi')^k}{(k!)}$$

we obtain

$$\Phi_n = \alpha^{-n} \sum_{k=n}^{\infty} k(k-1)\ldots(k-n+1)\alpha^{2k} \frac{\xi^{k-n}\xi'^k}{(k!)^2}$$

Setting $k = n + m$, we highlight the nth order *modified Bessel function* of the first kind I_n as follows inside the brackets:

$$\Phi_n = \xi'^n \sum_{m=0}^{\infty} \alpha^{2m+n} \frac{(\xi\xi')^m}{m!(m+n)!} = \frac{\xi'^n}{(\xi\xi')^{n/2}} \left[\sum_{m=0}^{\infty} \frac{(2\alpha\sqrt{\xi\xi'})^{n+2m}}{2^{n+m} m!(m+n)!} \right]$$

which leads to

$$\Phi_n = \left(\frac{\xi'}{\xi}\right)^{n/2} I_n(2\alpha\sqrt{\xi\xi'}) = \left(\frac{t-\frac{z}{v}}{t+\frac{z}{v}}\right)^{n/2} I_n\left(\alpha\sqrt{t^2-\frac{z^2}{v^2}}\right)$$

It remains to check the relation $u = u(z = 0, t)$ at the source level of the waveguide along positive and increasing z.

In order to do this, we have to expand $u = u(z = 0, t)$ into the base made up of $\Phi_n(z = 0, t)$, with coefficients chosen so that the expansion becomes a Taylor expansion $\sum_{n=0}^{\infty}(a_n/n!)t^n$ when α converges towards zero (lossless lines):

$$u(z=0,t) = \sum_{n=0}^{\infty} a_n \alpha^{-n} I_n(\alpha t)$$

So, the state of the *incident lossless guided wave* at any point set at a distance z from the source is given by the formula

$$u(z,t) = \sum_{n=0}^{\infty} a_n \alpha^{-n} \left(\frac{t-\frac{z}{v}}{t+\frac{z}{v}}\right)^{n/2} I_n\left(\alpha\sqrt{t^2-\frac{z^2}{v^2}}\right)$$

Since α is equal to the pure imaginary number $j\beta\sqrt{-1}$, the functions I_n become the classical *Bessel functions* J_n. In the case of TEM mode of propagation, the previous expansion becomes the Taylor expansion

$$\sum_{n=0}^{\infty} \frac{a_n}{n!}\left(t-\frac{z}{v}\right)^n$$

of an incident wave.

8.2.1.4. *Reverse wavelets*

Let us now deal with the reverse wavelets Φ'_n so that

$$\alpha\Phi'_n = G'\Phi'_{n-1}$$

This leads to

$$\Phi'_n = \frac{1}{\alpha}\frac{\partial \Phi'_{n-1}}{\partial \xi'} = \alpha^{-n}\frac{\partial^n \Phi_0}{\partial \xi'^n}$$

Then, we are led to exchange ξ and ξ' in the formula of the direct wavelets.

Coming back to the variables z and t, it happens that it is the result obtained by transforming v into $(-v)$: this highlights a *reflection* of incident waves and confirms the previous interpretation of the *reverse wavelets* as direct wavelets having a negative index.

Returning again to the TEM mode, α being zero, the direct wavelet Φ_n expansion and the reverse wavelet Φ'_n expansion become respectively the Taylor expansions of an incident wave depending on $(t - z/v)$ and of a reflected wave depending on $(t + z/v)$ as previously shown for lossless propagation according to the TEM mode.

8.2.2. *Application to the telegrapher's equation*

We recall the telegrapher's equation:

$$\left(v^2\frac{\partial^2}{\partial z^2} - \frac{\partial^2}{\partial t^2} - 2\alpha\frac{\partial}{\partial t}\right)u = 0 \qquad [8.12]$$

where α is a positive number. We can directly use the previous results for the guided waves as follows.

By means of the transformation $u = u_1 e^{-\alpha t}$, we obtain

$$\frac{\partial u}{\partial t} = \frac{\partial u_1}{\partial t}e^{-\alpha t} - \alpha e^{-\alpha t}u_1$$

$$\frac{\partial^2 u}{\partial t^2} = \frac{\partial^2 u_1}{\partial t^2}e^{-\alpha t} - 2\alpha e^{-\alpha t}\frac{\partial u_1}{\partial t} + \alpha^2 e^{-\alpha t}u_1$$

and

$$\left(v^2\frac{\partial^2}{\partial z^2} - \frac{\partial^2}{\partial t^2} - 2\alpha\frac{\partial}{\partial t}\right)u = \left(v^2\frac{\partial^2}{\partial z^2} - \frac{\partial^2}{\partial t^2} + \alpha^2\right)u_1 e^{-\alpha t}$$

then

$$\left(\frac{\partial^2}{\partial t^2} - v^2 \frac{\partial^2}{\partial z^2}\right) u_1 = \alpha^2 u$$

which is similar to the guided waves equation of which the solutions were expanded in wavelets previously.

So, we can write the "seed":

$$\Psi_0 = e^{-\alpha t} I_0\left(\alpha\sqrt{t^2 - \frac{z^2}{v^2}}\right)$$

The following are the direct and reverse generators:

$$G_\alpha = G + \alpha$$

$$G'_\alpha = G' + \alpha$$

so the telegrapher's equation becomes

$$A_\alpha u = \alpha^2 u \quad \text{with} \quad A_\alpha = G_\alpha G'_\alpha$$

We can check the relation: $\Psi_n = \Phi_n e^{-\alpha t}$. We have to obtain

$$\begin{cases} \Psi_n = G_\alpha \Psi_{n-1} = G\Psi_{n-1} + \alpha \Psi_{n-1} \\ \Psi_{n-1} = \Phi_{n-1} e^{-\alpha t} \end{cases}$$

but $t = \xi + \xi'$ and $G = \partial/\partial\xi$, so yielding

$$G\Psi_{n-1} = \frac{\partial}{\partial\xi}\left[e^{-\alpha(\xi+\xi')}\Phi_{n-1}\right] = -\alpha\Psi_{n-1} + e^{-\alpha(\xi+\xi')}G\Phi_{n-1}$$

therefore

$$\alpha\Psi_n = e^{-\alpha(\xi+\xi')}G\Phi_{n-1}$$

and

$$\alpha \Phi_n = G \Phi_{n-1}$$

as in the previous section.

So, the wavelet expansion is obtained directly as follows:

$$u(z,t) = \sum_{n=0}^{\infty} a_n \alpha^{-n} \left(\frac{t - \frac{z}{v}}{t + \frac{z}{v}} \right)^{n/2} I_n\left(\alpha \sqrt{t^2 - \frac{z^2}{v^2}} \right) e^{-\alpha t} \qquad [8.13]$$

where the coefficients a_n and α^{-n} are defined with respect to the knowledge of the source at the abscissa $z = 0$:

$$u(0,t) = \sum_{n=0}^{\infty} a_n \underbrace{\alpha^{-n} I_n(\alpha t) e^{-\alpha t}}_{\Psi_n(0,t)}$$

The wavelets at the source level are shown in Figure 8.1. Parameter α is a real positive number and the functions I_n are monotonically increasing.

However, for all $z > 0$:

$$\frac{t - \frac{z}{v}}{t + \frac{z}{v}} < 1 \quad \text{and} \quad \sqrt{t^2 - \frac{z^2}{v^2}} < t$$

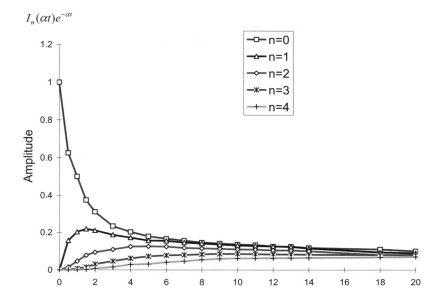

Figure 8.1. *Low order wavelets related to telegrapher's equation*

So, the convergence of $u(0, t)$ leads to that of $u(z, t)$. Physically, that means, without any reflected wave, the signal becomes increasingly flat during its propagation far away from the source.

8.2.3. *Convergence of wavelet expansions, numerical approximation*

8.2.3.1. *Space of signals expanded into wavelets*

In order to analyze the convergence of $u(0, t)$, it is necessary to define the functional space devoted to the wavelets. The asymptotic behavior of the Bessel functions when t increases indefinitely:

$$t \to \infty: \quad I_n(\alpha t) \cong \frac{e^{\alpha t}}{\sqrt{2\pi\alpha t}}$$

shows the functions expandable into wavelets have to decrease as $1/\sqrt{t}$ for large values of t. These functions u belong to the functional space $L^p([0,\infty])$, a set of functions having their power p absolutely integrable for $p > 2$: there are no finite energy signals. So, it is possible to represent by means of wavelets signals having a finite duration τ.

The wavelets do not make a system of orthogonal functions; their space is not fed with a Hilbertian norm related to a scalar product. The wavelets $I_n(\alpha t)e^{-\alpha t}$ are positive and bounded for $n \geq 0$.

Indeed, $\forall t \in [0, \infty[: I_n(\alpha t) < I_0(\alpha t)$ for $n \geq 0$, then $I_0(\alpha t)e^{-\alpha t}$ having a negative differential

$$\frac{d}{dt}\left[I_0(\alpha t)e^{-\alpha t}\right] = \alpha[I_0'(\alpha t) - I_0(\alpha t)]e^{-\alpha t}$$
$$= \alpha[I_1(\alpha t) - I_0(\alpha t)]e^{-\alpha t} < 0 \quad \text{because} \quad I_1(\alpha t) < I_0(\alpha t)$$

decreases from $t = 0$ where it is equal to 1: then it is positive and less than 1 as is $I_n(\alpha t)e^{-\alpha t}$.

Therefore, we propose using the space $L^\infty([0, \tau])$.

8.2.3.2. *General criterion of convergence*

We apply relation [8.9] as follows:

$$\|u - P_n u\|_{L^\infty([0,\tau])} \leq \|\Psi_0\|_{L^\infty([0,\tau])} \|G^n\| \sum_{m=0}^{\infty} |a_{m+n}| \|G^m\|$$

with the particular case $\Psi_m = I_m(\alpha t)e^{-\alpha t}$.

For $m = 0$, we see that

$$I_0(\alpha t)e^{-\alpha t} < 1 \quad \text{with} \quad \|\Psi_0\|_{L^\infty([0,\tau])} = 1$$

From the definition of G we have $\Psi_m = G^m \Psi_0$. Let us find an upper value of $\|G^m\|$. According to the theory of Bessel functions, we have

$$I_m(\alpha t) = \frac{\alpha t}{2m}[I_{m-1}(\alpha t) - I_{m+1}(\alpha t)]$$

From $I_{m+1}(\alpha t) \geq 0$ and with $t \leq \tau$, we obtain the inequalities

$$I_m(\alpha t) \leq \frac{\alpha t}{2m}I_{m-1}(\alpha t) \leq \frac{(\alpha t/2)^m}{m!}I_0(\alpha t) \leq \frac{(\alpha \tau/2)^m}{m!}I_0(\alpha t)$$

After multiplication by $\alpha^{-m}e^{-\alpha t}$ of both parts of the above inequalities and using the norm in $L^\infty([0,\tau])$, we obtain

$$\|\Psi_m\|_{L^\infty([0,\tau])} \leq \left\|\frac{(\alpha\tau/2)^m}{m!}\Psi_0\right\|_{L^\infty([0,\tau])} = \frac{(\alpha\tau/2)^m}{m!}\cdot 1$$

$\Psi_m = G^m \Psi_0$ leads to

$$\|G^m \Psi_0\|_{L^\infty([0,\tau])} \leq \frac{(\alpha\tau/2)^m}{m!}\cdot 1$$

from which we obtain the norm of the operator G^m:

$$\|G^m\| = \sup\frac{\|G^m\Psi_0\|_{L^\infty([0,\tau])}}{\|\Psi_0\|_{L^\infty([0,\tau])}} = \frac{(\alpha\tau/2)^m}{m!}$$

Using this result in relation [8.9], we obtain the convergence condition:

$$\|u - P_n u\|_{L^\infty([0,\tau])} \leq \frac{(\alpha\tau/2)^n}{n!}\sum_{m=0}^\infty |a_{m+n}|\alpha^{-(m+n)}\frac{(\alpha\tau/2)^m}{m!} < \varepsilon \qquad [8.14]$$

8.2.3.3. *Expansion of digital signals*

This result has to be developed again for applications. Indeed, it is met in applications of signals (Figure 8.2) in digital electronics.

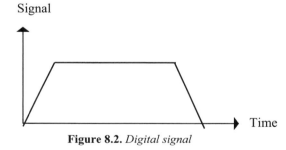

Figure 8.2. *Digital signal*

Therefore, it is necessary to expand linear or more generally polynomial slopes into series of wavelets $\alpha^{-n}I_n(\alpha t)e^{-\alpha t}$.

Let us show the procedure by means of the following *formula generating the Bessel functions* I_n:

$$e^{zchw} = \sum_{n=-\infty}^{+\infty} e^{nw} I_n(z) = I_0(z) + 2\sum_{n=1}^{\infty} ch(nw) I_n(z)$$

Setting $\chi = $ ch w, we use the *Tchebychev functions*:

$$T_n(\chi) = ch(n \arg ch \chi) \quad \text{with} \quad |\chi| \geq 1$$

where T_n is a polynomial having degree n. Thus

$$e^{\chi z} = I_0(z) + 2\sum_{n=1}^{\infty} T_n(\chi) I_n(z) \qquad [8.15]$$

This last formula is the basic one for applications in digital electronics.

In the case of a step $\chi = 1$, then

$$T_n(1) = ch(n \arg ch\ 1) = ch(n0) = 1, \quad z = \alpha t$$

After multiplication by $e^{-\alpha t}$ of the parts of equation [8.15], we get the required expansion:

$$1 = I_0(\alpha t)e^{-\alpha t} + 2I_1(\alpha t)e^{-\alpha t} + \cdots + 2I_n(\alpha t)e^{-\alpha t} + \cdots$$

In the case of a slope 1, let us differentiate [8.15], with respect to χ, and let χ converge towards 1:

$$T'_n(1) = \lim_{\chi \to 1} \left[\frac{n\ sh(n \arg ch \chi)}{\sqrt{1-\chi^2}} \right] = \lim_{\eta \to 1} \left(\frac{sh\ n\eta}{sh\ \eta} \right) = n^2$$

After multiplication of [8.15] after differentiation by $e^{-\alpha t}$ with $z = \alpha t$, we obtain

$$t = 2\left[\alpha^{-1} I_1(\alpha t)e^{-\alpha t}\right] + 8\alpha\left[\alpha^{-2} I_2(\alpha t)e^{-\alpha t}\right] + \cdots + 2n\alpha^{n-1}\left[\alpha^{-n} I_n(\alpha t)e^{-\alpha t}\right] + \cdots$$

$$[8.16]$$

More generally, $T_n^{(r)}(1)$ being zero for $n < r$, we obtain

$$t^r = 2\alpha^{-r} \sum_{n=r}^{\infty} T_n^{(r)}(1) I_n(\alpha t) e^{-\alpha t} \qquad [8.17]$$

This leads to a representation of any function expandable in a Taylor series.

8.2.3.4. *Convergence criterion for digital signals*

We now return to [8.14]. It is useful to know the behavior of the coefficients a_n in the wavelet expansion as n becomes increasingly large in the case of functions $f(t)$ defined by an integer r and a real number c. In this case $t^{-r}|f(t)|$ is bounded or otherwise defined by an integer r and a real number c so that $\forall t, |f(t)| < ct^r$ ("polynomial increasing functions").

Indeed, the Tchebychev functions $T_n(\chi)$ are solutions of the following differential equation:

$$(\chi^2 - 1)T_n''(\chi) + \chi T_n'(\chi) - n^2 T_n(\chi) = 0$$

Differentiating $(r-1)$ times with respect to χ applying the Leibniz formula, we obtain

$$(\chi^2 - 1)T_n^{(r+1)}(\chi) + \chi(2r - 1)T_n^{(r)}(\chi) + \left[(r-1)^2 - n^2\right]T_n^{(r-1)}(\chi) = 0 \qquad r > n$$

The value $\chi = 1$ leads to the recursive relation

$$T_n^{(r)}(1) = \frac{n^2 - (r-1)^2}{2r - 1} T_n^{(r-1)}(1) = \prod_{l=0}^{l=r-1} \frac{n^2 - 1^2}{2(1 + 1/2)} \qquad r > n \qquad [8.18]$$

because $T_n^{(0)}(1) = T_n(1) = 1$, which highlights that $T_n(r)(1)$ is a polynomial in n^2 having the degree r.

So, in applications concerning polynomial increasing signals, we need wavelet expansions having their coefficients $(a_n \alpha^{-n})$ belonging to the set of bounded $\{a_n \alpha^{-n} / n^q\}$, q being a positive integer.

In [8.14], we write

$$a_{m+n} = \frac{a_{m+n}}{(m+n)^q}(m+n)^q$$

We look for an upper value of $(m+n)^q/m!$ $(m \geq q)$. Let

$$m! = m(m-1)\ldots(m-q+1)(m-q)!$$

$$(m+n)^q < (m+n+1)(m+n+2)\ldots(m+n+1+q)$$

then

$$\frac{(m+n)^q}{m!} < \frac{1}{(m-q)!}\prod_{k=0}^{k=q}\left(1+\frac{n+1+q}{m-k}\right) < \frac{(n+q+2)^q}{(m-q)!}$$

Setting this result into [8.14], we have to split its second part into two; the first one with $m < q$ has no modification while the second one corresponding to $m \geq q$ is processed according to the last inequality about $(m+n)^q/m!$, thus yielding

$$\|u - P_n u\|_{L^\infty([0,\tau])} < \frac{(\alpha\tau/2)^n}{n!}\sum_{m=0}^{m=q}|a_{m+n}|\alpha^{-(m+n)}\frac{(\alpha\tau/2)^m}{m!}$$

$$+\frac{(n+q+2)^q}{n!}\left(\frac{\alpha\tau}{2}\right)^{n+q}\sum_{m=q}^{\infty}\frac{a_{m+n}\alpha^{-(m+n)}}{(m+n)^q}\frac{(\alpha\tau/2)^{m-q}}{(m-q)!} \qquad [8.19]$$

Furthermore, the inequality about $(m+n)^q/m!$ yields

$$\frac{(n+q+2)^q}{n!} < \frac{(2q+4)^q}{(n-q)!}$$

which leads to the final usable result corresponding to the case of bounded $\{a_n \alpha^{-n}/n^q\}$ by a number M:

$n > q:\ \|u - p_k u\|_{L^\infty([0,\tau])}$

$$< M\left(\frac{\alpha\tau}{2}\right)^q\left\{\sum_{m=0}^{m=q}(m+n)^q\frac{(\alpha\tau/2)^m}{m!} + (2q+4)^q\left(\frac{\alpha\tau}{2}\right)^q\left(e^{\alpha t/2}-1\right)\right\}\frac{(\alpha\tau/2)^{n-q}}{(n-q)!}$$

The general relation defining the dimension of the wavelet subspace.

Let us apply inequality [8.19] for a signal like t^r. If a_n is equal to $2\alpha^{n-r}T_n^{(r)}(1)$ according to [8.17] with $n > r$, recursive relation [8.18] yields

$$\frac{a_n}{n^{2r}}\alpha^{-n} = 2\alpha^{-r}\prod_{l=0}^{l=r-1}\frac{1-(1/n)^2}{2(1+1/2)} < 2\alpha^{-r}\frac{1}{2^{r-1}(r-1)!} \quad \text{since} \quad 1 < r < n$$

In the case of bounded $\{a_n \alpha^{-n}/n^q\}$ with $q = 2r$, [8.19] becomes $(n > 2r)$

$$\left\|t^r - P_n t^r\right\|_{L^\infty([0,\tau])} < \frac{(\alpha\tau/2)^{2r}}{2^{r-2}\alpha^r(r-1)!}$$

$$\times \left\{\left[\sum_{m=0}^{m=2r}(m+n)^{2r}\frac{(\alpha\tau/2)^m}{m!}\right] + 16(r+1)^{2r}\left(\frac{\alpha\tau}{2}\right)^{2r}\left(e^{\alpha t/2}-1\right)\right\}\frac{(\alpha\tau/2)^{n-2r}}{(n-2r)!} < 2\varepsilon$$

[8.20]

The stopping criterion of the wavelet expansion is defined by choosing $\varepsilon = 10^{-2}$ for instance.

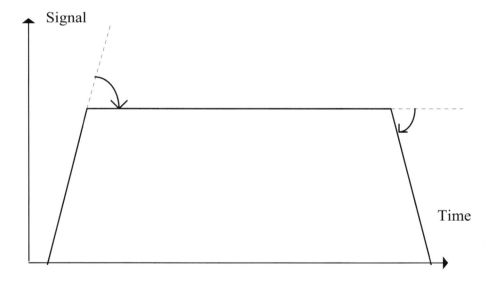

Figure 8.3. *Input signal shape*

8.2.3.5. *Digital signals having linear slopes*

In applications in digital electronics, it is usual to meet the signal depicted in Figure 8.3 that is made of three slopes.

Then, we need only $r = 1$ in this case with a duration $\tau = T$. So, we obtain

$$\left\| t^r - P_n t^r \right\|_{L^\infty([0,\tau])} < \frac{\alpha T^2}{2} \left[n^2 + (n+1)^2 \frac{\alpha T}{2} + (n+2)^2 \frac{(\alpha T)^2}{8} + 16(\alpha T)^2 \left(e^{\alpha t/2} - 1 \right) \right]$$

$$\times \frac{(\alpha T/2)^{n-2}}{(n-2)!} < 2\varepsilon$$

In order to use *Poisson's law* $\{\alpha^k/k!\}$ in applications, we propose a stopping criterion which replaces the last inequality when $n > 4$.

In order to do this, let us propose upper values of n^2, $(n+1)^2$, and $(n+2)^2$ as follows:

$$n^2 = (n-3+3)^2 = (n-3)^2 + 6(n-3) + 9 < (n-2)(n-3) + 6(n-2) + 9$$
$$< 16(n-2)(n-3)$$

$$(n+1)^2 = (n-3+4)^2 = (n-3)^2 + 8(n-3) + 16 < 25(n-2)(n-3)$$

$$(n+2)^2 = (n-3+5)^2 = (n-3)^2 + 10(n-3) + 25 < 36(n-2)(n-3)$$

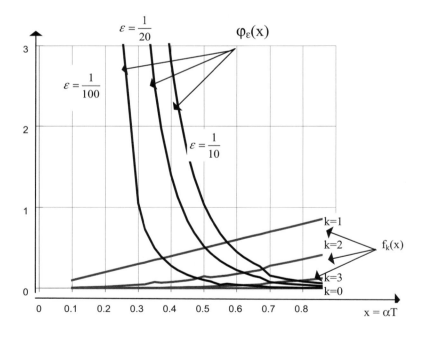

Figure 8.4. *Solution for $n \geq 4$*

Setting them into [8.20] leads to

$$\|t - P_n t\|_{L^\infty([0,\tau])} < \underbrace{\frac{\alpha^3 T^4}{8}\left[16 + \frac{25}{2}\alpha T + (\alpha T)^2\left(16 e^{\alpha t/2} - \frac{23}{2}\right)\right]}_{F(\alpha, t)} \frac{(\alpha T/2)^{n-4}}{(n-4)!} < 2\varepsilon \quad [8.21]$$

If data α and T give already $F(\alpha, T) < \varepsilon$, then n is less than 4 and we have to try $n = 3$ and $n = 2$ in [8.21].

In contrast, if $n > 4$, then we set $\xi = \alpha t$ and $x = \alpha T$.

Multiplying the parts of [8.21] by α, and replacing $\alpha \varepsilon$ by ε, the set of curves $f_k(x) = x^k / k!$ is built with $k = 0, 1, 2\ldots$, which leads to plotting the graphs (Figure 8.4) of the functions

$$\varphi_\varepsilon(x) = \frac{\varepsilon}{x^4 \left[1 + \frac{25}{32}x + x^2\left(e^{x/2} - \frac{23}{32}\right)\right]} \quad [8.22]$$

The intersections with the graphs of $f_k(x)$ give the value of $k = n - 4$ corresponding to any given αT.

8.3. Conclusion

The requirement to be close to the direct time domain method, without going into the frequency domain and coming back to the time domain, has led us to develop a wavelet theory tied to solving linear partial differential equations, which should not be mistaken for those defined by Meyer in the context of signal processing.

The method becomes increasingly more efficient than the frequency method as the transition times of digital signals become shorter and shorter.

Chapter 9

Applications of the Wavelet Method

The development of the previous chapter is extended to lossy coupled lines by means of the spectral theory of matrices. It is completed by a multiple reflection analysis in the *space of wavelets*, so leading to an extension of the recursive matrix equation for coupled lossless lines to coupled lossy lines directly in the time domain by means of a group structure. This is achieved by whole complex lossy network modeling based on a *time-domain wavelet matrix state equation*, the computation and memory requirements of which are compared to those needed by the frequency method.

An extension of the wavelets method is proposed in the form of a 3D electromagnetic perturbation analysis.

9.1. Coupled lossy transmission lines in the TEM approximation

9.1.1. *Wavelets in homogenously coupled lossy transmission lines*

9.1.1.1. Recalling the homogenously coupled lossy transmission line matrix equation

The matrix blocks [M] and [N] have been defined in Chapter 7 devoted to coupled lossy lines, the equation of which was written as

$$v\frac{\partial}{\partial z}\begin{bmatrix}[V]\\[J]\end{bmatrix}+[M]\frac{\partial}{\partial t}\begin{bmatrix}[V]\\[J]\end{bmatrix}+[N]\begin{bmatrix}[V]\\[J]\end{bmatrix}=\begin{bmatrix}[0]\\[0]\end{bmatrix}$$

where [V] and [J] are the voltage and the current vectors, respectively.

9.1.1.2. *Lossy wave propagation matrix operator*

Now, first we have to find the second-order equation arising from the equation above by means of the wavelets method. Let $[T_+]$ be the following operator devoted to "direct waves":

$$[T_+] = [1]v\frac{\partial}{\partial z} + [M]\frac{\partial}{\partial t} + [N] \quad\quad [9.1]$$

We define another operator devoted to "reverse waves":

$$[T_-] = [1]v\frac{\partial}{\partial z} - [M]\frac{\partial}{\partial t} - [N] \quad\quad [9.2]$$

Let us compute the product

$$[T] = [T_+][T_-] = [T_-][T_+] \quad\quad [9.3]$$

$$\begin{aligned}[] [T_+][T_-] &= [1]\left(v\frac{\partial}{\partial z}\right)^2 - \left\{[M]\frac{\partial}{\partial t}+[N]\right\}^2 \\ &= [1]v^2\frac{\partial^2}{\partial z^2} - [M]^2\frac{\partial^2}{\partial t^2} - \{[M][N]+[N][M]\}\frac{\partial}{\partial t} - [N]^2 \end{aligned}$$

$$[M][N] = \begin{bmatrix} \dfrac{\sigma}{\varepsilon v}[1] & [0] \\ [0] & \dfrac{\sigma}{\varepsilon v}[1] \end{bmatrix} v$$

$$[N][M] = \begin{bmatrix} R[Z_{c0}]^{-1} & [0] \\ [0] & \dfrac{\sigma}{\varepsilon v}[1] \end{bmatrix} v$$

Let $[[\alpha]]$ be the matrix block

$$[\alpha]\begin{bmatrix} [1] & [0] \\ [0] & [1] \end{bmatrix} = \begin{bmatrix} [\alpha] & [0] \\ [0] & [\alpha] \end{bmatrix}$$

Defining the matrix

$$[\alpha] = \frac{1}{2}\left\{\frac{\sigma}{\varepsilon}[1] + Rv[Z_{c0}]^{-1}\right\}$$

we obtain

$$[N]^2 = \begin{bmatrix} \left[\frac{\sigma}{\varepsilon}Rv[Z_{c0}]^{-1}\right] & [0] \\ [0] & \left[\frac{\sigma}{\varepsilon}Rv[Z_{c0}]^{-1}\right] \end{bmatrix} = \left\{2\frac{\sigma}{\varepsilon}[[\alpha]] - \frac{\sigma^2}{\varepsilon^2}[1]\right\}$$

where, in order to simplify the notation, the matrix block

$$\begin{bmatrix} [1] & [0] \\ [0] & [1] \end{bmatrix}$$

is written as [1]. Also

$$[M][N] + [N][M] = 2[[\alpha]] \qquad [9.4]$$

Then, we obtain the lossy wave propagation matrix operator:

$$[T] = [1]\left(v^2 \frac{\partial^2}{\partial z^2} - \frac{\partial^2}{\partial t^2}\right) - 2[[\alpha]]\frac{\partial}{\partial t} - 2\frac{\sigma}{\varepsilon}[[\alpha]] + 2\frac{\sigma^2}{\varepsilon^2}[1] \qquad [9.5]$$

9.1.1.3. *Defining the matrix operator [A] and the matrix [Λ] so that [A][u] = [Λ][u]*

The equation $[T][u] = [0]$ is not rigorously a telegrapher's equation. The operator $[T]$ can be developed as

$$[T] = \left\{[[\alpha]] - \frac{\sigma}{\varepsilon}[1]\right\}^2 - [A]$$

where

$$[A] = \left\{[1]\frac{\partial}{\partial t} + [[\alpha]]\right\}^2 - [1]v^2 \frac{\partial^2}{\partial z^2}$$

246   Electromagnetism and Interconnections

so that the equation $[T][u] = [0]$ becomes $[A][u] = [\Lambda][u]$ with

$$[A] = \left\{[[\alpha]] - \frac{\sigma}{\varepsilon}[1]\right\}^2 = \left\{\frac{1}{2}[Rv[Z_{c0}]^{-1}] - \frac{1}{2}\frac{\sigma}{\varepsilon}[1]\right\}^2$$

being a real symmetric matrix, which is "positive" because its eigenvalues, equal to $(1/4)\{\text{eigenvalue}[[\alpha]] - (\sigma/\varepsilon)\}$, are positive.

9.1.1.4. *Diagonalization of the matrix operator* [A]

Operator [A] and matrix [Λ], depending on [[α]] only, commute and have the same eigenvectors: therefore, they can be diagonalized simultaneously. So, the matrix equation $[A][u] = [\Lambda][u]$ can be shared into as many "scalar" equations as there are lossy coupled lines.

Then, each "scalar" equation gives a family of direct or reverse wavelets $\Psi_n(\lambda)$ that are computed by means of the spectral theory of matrices developed above.

Let us write the spectral expansion of the matrix [Λ]: $[\Lambda] = \sum_i \lambda_i [P_i]$, where λ_i are the different eigenvalues of [Λ] and [P_i] are the projectors corresponding to the eigen-subspaces. This leads to the spectral expansion of the matrix function [$\Psi_n([\Lambda])$]:

$$[\Psi_n([\Lambda])] = \sum_i \Psi_n(\lambda_i)[P_i]$$

Applying the operator [A], we obtain

$$[A][\Psi_n([\Lambda])] = [A]\sum_i \Psi_n(\lambda_i)[P_i] = \sum_i [A]\Psi_n(\lambda_i)[P_i]$$

Knowing matrix [Λ] is positive, we can write

$$[\Lambda]^{1/2} = \sum_i \lambda_i^{1/2}[P_i] \quad \text{because} \quad \lambda_i \geq 0 \;\forall i$$

Due to this, [A] becomes

$$\left\{[1]\frac{\partial}{\partial t} + [1]\frac{\sigma}{\varepsilon} + [\Lambda]^{1/2}\right\}^2 - [1]v^2 \frac{\partial^2}{\partial z^2}$$

and has the following spectral expansion which defines the required diagonalization of [A]:

$$[A] = \sum_k A(\lambda_k)[P_k]$$

where

$$A(\lambda_k) = \left(\frac{\partial}{\partial t} + \frac{\sigma}{\varepsilon} + \lambda_k^{1/2}\right)^2 - v^2 \frac{\partial^2}{\partial z^2}$$

is a "scalar" operator.

This "scalar" operator has the wavelets $\Psi_n(\lambda_k)$ corresponding to

$$A(\lambda_k)\Psi_n(\lambda_k) = \lambda_k \Psi_n(\lambda_k) \quad \forall n \in N,\ k \in N$$

9.1.1.5. *Direct and reverse matrix wavelets onto an infinite transmission line*

Recalling that the projectors $[P_k]$ have the properties $[P_k][P_i] = 0$ if $k \neq i$ and $= [P_i]$ if $k = i$, we obtain

$$[A][\Psi_n([\Lambda])] = \sum_{i,k} A(\lambda_k)\Psi_n(\lambda_i)[P_k][P_i] = \sum_i A(\lambda_i)\Psi_n(\lambda_i)[P_i]$$

$$= \sum_i \lambda_i \Psi_n(\lambda_i)[P_i] = \left\{\sum_i \lambda_i[P_i]\right\}\left\{\sum_j \Psi_n(\lambda_j)[P_j]\right\} = [\Lambda]\Psi_n([\Lambda])$$

This leads directly to the solutions $[u]$ of the equation $[T][u] = [0]$ according to

$$[u] = [\Psi][a]$$

where $[\Psi]$ is a square matrix and $[a]$ is a constant vector. These solutions are directly obtained by setting the matrix $[\Lambda]$ instead of λ into the function $\Psi(\lambda)$ solution of $A(\lambda)\Psi = \lambda\Psi$.

This last equation becomes the telegrapher's equation by means of the transformation $\Psi = \Omega e^{-(\sigma/\varepsilon)t}$.

Indeed,

$$\left(\frac{\partial}{\partial t}+\frac{\sigma}{\varepsilon}+\lambda^{1/2}\right)\left(\Omega e^{-(\sigma/\varepsilon)t}\right) = e^{-(\sigma/\varepsilon)t}\left(\frac{\partial}{\partial t}+\lambda^{1/2}\right)(\Omega)$$

$$\left(\frac{\partial}{\partial t}+\frac{\sigma}{\varepsilon}+\lambda^{1/2}\right)^2\left(\Omega e^{-(\sigma/\varepsilon)t}\right) = e^{-(\sigma/\varepsilon)t}\left(\frac{\partial}{\partial t}+\lambda^{1/2}\right)^2(\Omega)$$

$$\begin{aligned}[][A(\lambda)-\lambda]\left(\Omega e^{-(\sigma/\varepsilon)t}\right) &= \left[\left(\frac{\partial}{\partial t}+\frac{\sigma}{\varepsilon}+\lambda^{1/2}\right)^2 - v^2\frac{\partial^2}{\partial z^2} - \lambda\right]\left(\Omega e^{-(\sigma/\varepsilon)t}\right) \\ &= e^{-(\sigma/\varepsilon)t}\left[\left(\frac{\partial}{\partial t}+\lambda^{1/2}\right)^2 - v^2\frac{\partial^2}{\partial z^2} - \lambda\right](\Omega) \\ &= e^{-(\sigma/\varepsilon)t}\left(\frac{\partial^2}{\partial t^2}+2\lambda^{1/2}\frac{\partial}{\partial t} - v^2\frac{\partial^2}{\partial z^2}\right)(\Omega)\end{aligned}$$

So, the equation $[A(\lambda)-\lambda]\Psi = 0$ becomes the telegrapher's equation:

$$\left(\frac{\partial^2}{\partial t^2}+2\lambda^{1/2}\frac{\partial}{\partial t} - v^2\frac{\partial^2}{\partial z^2}\right)(\Omega) = 0$$

the wavelet solution of which was given in Chapter 8 in the form of [8.13].

Carrying out the substitution $\lambda \to [\Lambda]$ in the corresponding scalar expressions, and returning to Ψ, we obtain directly the following matrix wavelets. *"Direct"* *wavelets*:

$$[\Psi_n([\Lambda])] = [\Lambda]^{-\frac{n}{2}}\left(\frac{t-\frac{z}{v}}{t+\frac{z}{v}}\right)^{n/2} I_n\left\{[\Lambda]^{1/2}\sqrt{t^2-\frac{z^2}{v^2}}\right\}\exp\left\{-[\Lambda]^{1/2}t\right\}e^{-(\sigma/\varepsilon)t} \qquad [9.6]$$

"Reverse" wavelets:

$$[\Psi'_n[\Lambda]] = [\Lambda]^{-\frac{n}{2}}\left(\frac{t+\frac{z}{v}}{t-\frac{z}{v}}\right)^{n/2} I_n\left\{[\Lambda]^{1/2}\sqrt{t^2-\frac{z^2}{v^2}}\right\}\exp\left\{-[\Lambda]^{1/2}t\right\}e^{-(\sigma/\varepsilon)t} \qquad [9.7]$$

Applications of the Wavelet Method 249

We note $[\Psi_n]$ as $[\Psi_n^+]$ solution of

$$[T^+][\Psi_n^+] = 0$$

and $[\Psi'_n]$ as $[\Psi_n^-]$ solution of

$$[T^-][\Psi_n^-] = 0$$

9.1.1.6. *Convergence and numerical approximation of matrix wavelet expansions*

It remains to analyze the convergence of the matrix wavelet expansion made up of $[\Psi_n]$ multiplied by the vector coefficient $[a_n]$ on its left-hand side so that a Taylor expansion can be obtained when the line becomes lossless:

$$\sum_{n=0}^{\infty} [\Psi_n][a_n]$$

In order to do that, it is necessary to consider the spectral expansion of $[\Psi_n]$ given above, and also the definition of the norm of matrices.

The matrix $[\Psi_n]$ depends on $[\Lambda]$ which is a positive symmetric real matrix; its highest eigenvalue is called the *spectral radius* of it and defines its norm. This norm, being equal to the so-called Euclidean norm for symmetric and real matrices, can be denoted as $\|[\Psi_n]\|_2$.

Concerning the vector $[a_n]$, we propose the classical Euclidean norm denoted as $\|[a_n]\|_2$.

Using these definitions, we apply the inequality

$$\|[\Psi_n][a_n]\|_2 \leq \|[\Psi_n]\|_2 \|[a_n]\|_2 \qquad [9.8]$$

that we complete by the triangle inequality in the first two parts:

$$\left\|\sum_{n=0}^{\infty} [\Psi_n][a_n]\right\|_2 \leq \sum_{n=0}^{\infty} \|[\Psi_n][a_n]\|_2 \qquad [9.9]$$

So, replacing a_n by $\|[a_n]\|_2$ in inequality [8.14], while α is replaced by the spectral radius of the positive symmetric real matrix $[\Lambda]^{1/2}$, we obtain the stopping criterion of the matrix wavelet expansion:

$$\left\| \|[u]-[P_n u]\|_2 \right\|_{L^\infty([0,\tau])} \le \frac{\left(\dfrac{\tau}{2}\right)^n}{n!} \sum_{m=0}^{\infty} \|[a_{m+n}]\|_2 \frac{\left(\dfrac{\tau}{2}\right)^m}{m!} < \varepsilon$$

[9.10]

To conclude, the wavelet expansion becomes a classical Taylor expansion

$$\sum_{n=0}^{\infty} \frac{\left(t-\dfrac{z}{v}\right)^n}{n!}[a_n]$$

when $\left\|[A]^{1/2}\right\|_2$ converges towards zero, so the coupled lines become lossless.

9.1.2. *Multiple reflections into lossy coupled lines*

9.1.2.1. *Generators of the wavelet base*

Let us begin with the equation $[A][u] = [A][u]$ from the previous discussion, where the operator

$$[A] = \left\{[1]\frac{\partial}{\partial t}+[[\alpha]]\right\}^2 - [1]v^2 \frac{\partial^2}{\partial z^2}$$

has to be factorized as regards obtaining "direct" and "reverse" wavelets able to account for multiple reflections.

So, we set

$$[A] = [G_{[\alpha]}][G'_{[\alpha]}]$$

with the "direct" generator:

$$[G_{[\alpha]}] = [1]\left(v\frac{\partial}{\partial z}+\frac{\partial}{\partial t}\right)+[[\alpha]]$$

and the "reverse" generator:

$$[G'_{[\alpha]}] = [1]\left(-v\frac{\partial}{\partial z}+\frac{\partial}{\partial t}\right)+[[\alpha]]$$

Then, we would like to find

$$\begin{cases} [A]^{1/2}[\Psi_n] = [G_{[\alpha]}][\Psi_{n-1}] \\ [A]^{1/2}[\Psi_n] = [G'_{[\alpha]}][\Psi_{n+1}] \end{cases}$$ [9.11]

recalling that

$$\begin{cases} [A] = \left\{ \frac{1}{2}\left[Rv[Z_{c0}]^{-1} \right] - \frac{1}{2}\frac{\sigma}{\varepsilon}[1] \right\}^2 \\ [[\alpha]] = \frac{1}{2} Rv[Z_{c0}]^{-1} + \frac{1}{2}\frac{\sigma}{\varepsilon}[1] \end{cases}$$

According to the spectral expansion of $[A]$ and of $[A]^{1/2}$, these matrices commute with $[G_{[\alpha]}]$ and $[G'_{[\alpha]}]$. Therefore relations [9.11] show that $[\Psi_n]$ are solutions of $[A][u] = [A][u]$.

Indeed

$$[A][\Psi_n] = [G_{[\alpha]}][A]^{1/2}[\Psi_{n-1}] = [A]^{1/2}[G_{[\alpha]}][\Psi_{n-1}] = [A][\Psi_n]$$

which yields

$$\begin{cases} v\frac{\partial}{\partial z}[\Psi_n] = \frac{1}{2}[A]^{1/2}\{[\Psi_{n+1}] - [\Psi_{n-1}]\} \\ \frac{\partial}{\partial t}[\Psi_n] = \frac{1}{2}[A]^{1/2}\{[\Psi_{n+1}] + [\Psi_{n-1}]\} - [[\alpha]][\Psi_n] \end{cases}$$ [9.12]

9.1.2.2. Derivation and translation matrix operators in the wavelet space

From now, we chain all the above matrix relations into a single one by means of the matrix $[\Psi]$ of which the transpose is

$$[\Psi]^\top = [[\Psi_0],[\Psi_1],\ldots[\Psi_n],\ldots]$$

so that relations [9.12] becomes a compact matrix relation:

$$\left.\begin{array}{l}2[\Lambda]^{-1/2}\dfrac{\partial}{\partial t}[\Psi]=[B][\Psi]\\[6pt] 2[\Lambda]^{-1/2}\dfrac{\partial}{\partial z}[\Psi]=[C][\Psi]\end{array}\right\} \qquad [9.13]$$

The known commutability of the differential operators $\partial/\partial t$ and $\partial/\partial z$ according to the Schwarz theorem [VAL2] demonstrates the commutability of matrices $[B]$ and $[C]$ then of any $[B]^p$ and $[C]^q$, p and q being integers, without any further computation.

Moreover, this commutability of matrices $[B]$ and $[C]$ can be applied in the derivations involved with the Taylor expansion:

$$[\Psi^+(z,t)] = \left[\Psi^+\left(0+z, t+\dfrac{z}{v}-\dfrac{z}{v}\right)\right]$$

$$= \sum_{m=0}^{\infty}\dfrac{1}{m!}\left(z\dfrac{\partial}{\partial z}-\dfrac{z}{v}\dfrac{\partial}{\partial t}\right)^m \left[\Psi^+\left(0, t+\dfrac{z}{v}\right)\right]$$

which defines the new matrix $[E_{[\Lambda]}(z)]$:

$$[\Psi^+(z,t)] = [E_{[\Lambda]}(z)]\left[\Psi^+\left(0, t+\dfrac{z}{v}\right)\right] \qquad [9.14]$$

Then, we find

$$[E_{[\Lambda]}(z)] = \sum_{m=0}^{\infty}\dfrac{1}{v^m}\dfrac{z^m}{2^m m!}[\Lambda]^{m/2}([C]-[B])^m = \exp\left\{\dfrac{z}{2v}[\Lambda]^{1/2}[[C]-[B]]\right\} \qquad [9.15]$$

The matrix $[E_{[\Lambda]}(z)]$ has the following expressions:

$$[E_{[\Lambda]}(z)] = \begin{bmatrix} [1] & \dfrac{z}{2v}[\Lambda]^{1/2} & \dfrac{1}{2!}\left\{\dfrac{z}{2v}[\Lambda]^{1/2}\right\}^2 & \dfrac{1}{3!}\left\{\dfrac{z}{2v}[\Lambda]^{1/2}\right\}^3 & \cdots \\[8pt] [0] & [1] & \dfrac{z}{2v}[\Lambda]^{1/2} & \dfrac{1}{2!}\left\{\dfrac{z}{2v}[\Lambda]^{1/2}\right\}^2 & \cdots \\[8pt] [0] & [0] & [1] & \dfrac{z}{2v}[\Lambda]^{1/2} & \cdots \\[8pt] [0] & [0] & [0] & [1] & \cdots \\[4pt] \vdots & \vdots & \vdots & \vdots & \ddots \end{bmatrix}$$

We note that at the source

$$[E_{[A]}(0)] = [1] \quad [9.16]$$

and in the case of a lossless media

$$[E_{[0]}(z)] = [1] \quad [9.17]$$

Accounting for the commutability of matrices $[B]$ and $[C]$, we get easily from [9.15] the interesting properties of $[E_{[A]}(z)]$ leading to a *group structure* in the space domain (while it is known in functional analysis that there is only a semi-group structure in the time domain) as follows:

$$\begin{aligned} [E_{[A]}(z+z')] &= [E_{[A]}(z)][E_{[A]}(z')] \\ [E_{[A]}(z)]^{-1} &= [E_{[A]}(-z)] \end{aligned} \quad [9.18]$$

9.1.2.3. *Direct and reverse matrix wavelets, recursive state equation*

Applying [9.18] and exchanging the "reverse" and "direct" wavelets by transforming z into $(-z)$, we obtain

$$[\Psi^-(z,t)] = [\Psi^+(-z,t)] = [E_{[A]}(-z)]\left[\Psi^+\left(0, t - \frac{z}{v}\right)\right] \quad [9.19]$$

Factorization [9.3] leads to the fact that any $[\Psi_n]$ can be shared into a "direct" wavelet $[\Psi_n^+]$ solution and a "reverse" wavelet $[\Psi_n^-]$ solution, respectively, of the equations

$$\begin{aligned} [T_+][\Psi_n^+] &= 0 \\ [T_-][\Psi_n^-] &= 0 \end{aligned} \quad [9.20]$$

which are new formulations of [9.2] and [9.1], respectively.

Summing [9.1] and [9.2], we obtain

$$[T_+] + [T_-] = 2[1]v\frac{\partial}{\partial z}$$

Applying this operator to $[\Psi_n] = [\Psi_n^+] + [\Psi_n^-]$ and being aware of [9.20], we obtain the final wavelet equation for lossy coupled lines:

$$2v\frac{\partial}{\partial z}[\Psi_n] = [T_+][\Psi_n^-] + [T_-][\Psi_n^+] \qquad [9.21]$$

Accounting for [9.2], [9.1], and [9.13], relation [9.20] becomes

$$\left.\begin{aligned}[\Lambda]^{1/2}[C][\Psi] &= [S_+][\Psi^-] + [S_-][\Psi^+] \\ [S_+] &= \frac{1}{2}[\Lambda]^{1/2}[C] - \frac{1}{2}[M][\Lambda]^{1/2}[B] - [N] \\ [S_-] &= \frac{1}{2}[\Lambda]^{1/2}[C] + \frac{1}{2}[M][\Lambda]^{1/2}[B] + [N]\end{aligned}\right\} \qquad [9.22]$$

To conclude, relations [9.14] and [9.19] yield

$$[\Lambda]^{1/2}[C][\Psi(z,t)] = [S_-][E_{[\Lambda]}(z)]\left[\Psi^+\left(0, t+\frac{z}{v}\right)\right] + [S_+][E_{[\Lambda]}(-z)]\left[\Psi^+\left(0, t-\frac{z}{v}\right)\right] \qquad [9.23]$$

where $(t + z/v)$ corresponds to the present time that requires the time shift $t \to t - z/v$,

$$\left[\Psi^+\left(0, t+\frac{z}{v}\right)\right]$$

corresponds to the incident signals coming from the origin of the transmission lines, $[\Psi(z,t)]$ corresponds to the signal sources, and

$$\left[\Psi^+\left(0, t-\frac{z}{v}\right)\right]$$

is related to the signals reflected at the line input.

Processing the time shift $t \to t - z/v$ and accounting for relation [9.17], we obtain the final result:

$$[A]^{1/2}[C]\left[\Psi\left(z,t-\frac{z}{v}\right)\right]-[S_-][E_{[A]}(z)][\Psi^+(0,t)]=[S_+][E_{[A]}(-z)]\left[\Psi^+\left(0,t-\frac{2z}{v}\right)\right]$$

[9.24]

So, we have demonstrated a unitary processing of both lossy and lossless lines directly in the time domain. The difference between lossy and lossless lines lies in the fact that lossy wave propagation requires an infinite dimension space, that of the "wavelets", and shows a disturbance of waveform shape tied to the infinite dimension matrix $[E_{[A]}]$.

9.1.3. Comparative analysis of frequency and wavelets methods

Let us deal with complex networks in the case of losses. For this, we need to model the lumped elements at the line terminals. This can be achieved by means of the wavelets method using the first of relations [9.13] which can be written as

$$\frac{\partial}{\partial t}[\Psi]=\frac{1}{2}[A]^{1/2}[B][\Psi]$$

Accounting for equation [9.24], we are able to obtain the new *matrix state equation* of the *state vector* $[X]$ as follows:

$$[A]\left[X\left([z],t-\frac{[z]}{v}\right)\right]=[B]\left[Y\left([z_0],t-\frac{[z_0]}{v}\right)\right]+[C]\left[X\left([z],t-\frac{2[z]}{v}\right)\right]$$

By means of the approximation into a subspace having dimension n, the computation requirement, related to the Gauss algorithm, is made up of $(4Ln)^3/3$ elementary operations like additions and multiplications while the memory capacity has to store $3(4Ln)^2+2(4Ln)$ numbers.

Appendix D shows a modified Gauss-Seidel algorithm which is a recursive method of solving [9.24] by saving a lot of storage capacity.

In order to consider the storage and computation requirements in comparison with the direct time-domain matrix tied to lossless lines, the matrix structure needed for modeling by means of wavelets, two coupled lossy lines having the same length with their input-output loads, is shown in Figure 9.1.

The number of rows and columns needed for wavelet modeling lines having very small losses is three times higher than that needed for modeling lossless lines, while the number becomes seven times higher in the case of lines having strong losses.

The wavelet method is better than the frequency method for modeling digital signals since $n < 4$ or $4 \leq n < 2\sqrt[3]{2N+1}$, where N is the dimension of the vector subspace needed by the numerical approximation of the discrete Fourier transform:

$$N = 0.19\varepsilon^{-1/3} T^{2/3} \left(\frac{ds}{dt}\right)^{2/3}$$

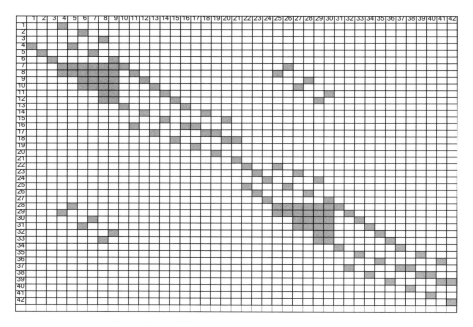

Figure 9.1. *Filling state of the [A] matrix with n = 3 (gray: non-zero elements; white: zero elements)*

9.2. Extension to 3D wavelets and electromagnetic perturbations

9.2.1. *Basic second-order partial differential equation of electromagnetic waves*

We now consider the propagation of electromagnetic waves in homogenous media around cylindrical lossless conductors. In this case, Maxwell's equations presented in Chapter 1

$$\vec{\nabla} \wedge \underline{\tilde{H}} = \sigma \vec{E} + \varepsilon \frac{\partial \vec{E}}{\partial t}$$

$$\vec{\nabla} \wedge \underline{\vec{E}} = -\sigma_m \underline{\tilde{H}} - \mu \frac{\partial \underline{\tilde{H}}}{\partial t}$$

lead to the second-order partial differential equations where we do not deal with the usual zero divergence of electrical and magnetic fields needed for the wavelets concept:

$$\vec{\nabla} \wedge (\vec{\nabla} \wedge \underline{\tilde{H}}) = -\left(\sigma + \varepsilon \frac{\partial}{\partial t}\right)\left(\sigma_m + \mu \frac{\partial}{\partial t}\right)\underline{\tilde{H}}$$

$$\vec{\nabla} \wedge (\vec{\nabla} \wedge \vec{E}) = -\left(\sigma_m + \mu \frac{\partial}{\partial t}\right)\left(\sigma + \varepsilon \frac{\partial}{\partial t}\right)\vec{E}$$

9.2.2. Obtaining the wavelet generating equation: $Au = \Lambda u$

In the following, we assume $\sigma_m = 0$. Letting $\vec{E} = \vec{E}' e^{-\sigma t/2\varepsilon}$, we obtain

$$\frac{\partial \vec{E}}{\partial t} = e^{-\sigma t/2\varepsilon}\left(\frac{\partial}{\partial t} - \frac{\sigma}{2\varepsilon}\right)\vec{E}' \quad \text{and} \quad \left(\sigma + \varepsilon \frac{\partial}{\partial t}\right)\vec{E} = e^{-\sigma t/2\varepsilon} \varepsilon \left(\frac{\partial}{\partial t} + \frac{\sigma}{2\varepsilon}\right)\vec{E}'$$

and similarly for the magnetic field. Putting these expressions into the above second-order equations and accounting for all the media parameters μ, ε, σ being constant, multiplying both sides of the equations by $e^{\sigma t/2\varepsilon}$ yields the equation

$$\vec{\nabla} \wedge (\vec{\nabla} \wedge \vec{E}') = -\mu\varepsilon \left(\frac{\partial}{\partial t} - \frac{\sigma}{2\varepsilon}\right)\left(\frac{\partial}{\partial t} + \frac{\sigma}{2\varepsilon}\right)\vec{E}'$$

This last equation is rewritten with regards to the required *wavelet generating equation* $Au = \Lambda u$ and the wave velocity $v = 1/\sqrt{\mu\varepsilon}$, so defining the operator A and the scalar Λ:

$$\underbrace{\left[\vec{\nabla} \wedge (\vec{\nabla} \wedge \cdot) + \frac{1}{v^2}\frac{\partial^2}{\partial t^2}\right]}_{A}\vec{E}' = \underbrace{\left(\frac{\mu\sigma^2}{4\varepsilon}\right)}_{\Lambda}\vec{E}'$$

9.2.3. Direct and reverse generators of the wavelet base

According to the equation $Au = \Lambda u$ defining the operator A, we deal with the commutative factorization of this operator which gives the direct generator G and the reverse generator G' as follows:

$$A = \underbrace{\left[\frac{1}{v}\frac{\partial}{\partial t} + j(\nabla \vee \cdot)\right]}_{G} \underbrace{\left[\frac{1}{v}\frac{\partial}{\partial t} - j(\nabla \vee \cdot)\right]}_{G'}$$

Then the wavelet family is created by means of the recursive relations

$$\alpha \Psi_{n+1} = G\Psi_n$$
$$\alpha \Psi_{n-1} = G'\Psi_n$$

where

$$\alpha = \Lambda^{1/2} = \sqrt{\frac{\mu \sigma^2}{4\varepsilon}}$$

has become a classical parameter in our analysis.

These last relations give the action of the differential operators on the wavelets, as has already been met:

$$\frac{\partial \Psi_n}{\partial t} = \frac{\alpha v}{2}(\Psi_{n-1} + \Psi_{n+1})$$

$$\nabla \wedge \Psi_n = j\frac{\alpha}{2}(\Psi_{n-1} - \Psi_{n+1})$$

There are wavelets denoted as $\Psi_n^{E'}$ related to the electrical field $\vec{E}'$ and other wavelets denoted as $\Psi_n^{H'}$ related to the magnetic field $\vec{\tilde{H}}'$. The wavelets related to the electrical and magnetic fields have to be tied by the same relations as these fields. The involved relations come directly from Maxwell's equations with $\vec{E} = \vec{E}'e^{-\sigma t/2\varepsilon}$ and similarly for the magnetic field $\vec{\tilde{H}} = \vec{\tilde{H}}'e^{-\sigma t/2\varepsilon}$. So, we obtain

$$\tilde{\vec{\nabla}} \wedge \tilde{\vec{H}}' e^{-\sigma t/2\varepsilon} = \sigma \underline{\vec{E}} + \varepsilon \frac{\partial \underline{\vec{E}}}{\partial t} = e^{-\sigma t/2\varepsilon} \varepsilon \left(\frac{\partial}{\partial t} + \frac{\sigma}{2\varepsilon} \right) \vec{E}'$$

$$\tilde{\vec{\nabla}} \wedge \vec{E}' e^{-\sigma t/2\varepsilon} = -\mu \frac{\partial \underline{\vec{H}}}{\partial t} = -\mu e^{-\sigma t/2\varepsilon} \varepsilon \left(\frac{\partial}{\partial t} + \frac{\sigma}{2\varepsilon} \right) \vec{H}'$$

Simplifying and replacing the fields by their corresponding wavelets yields the modified Maxwell's equations

$$\tilde{\vec{\nabla}} \wedge \Psi_n^{H'} = \varepsilon \left(\frac{\partial}{\partial t} + \frac{\sigma}{2\varepsilon} \right) \Psi_n^{E'}$$

$$\tilde{\vec{\nabla}} \wedge \Psi_n^{E'} = -\mu \left(\frac{\partial}{\partial t} - \frac{\sigma}{2\varepsilon} \right) \Psi_n^{H'}$$

Dealing with the recursive equations $\alpha \Psi_{n+1}^{E'} = G \Psi_n^{E'}$ and $\alpha \Psi_{n+1}^{H'} = G \Psi_n^{H'}$, with

$$G = \frac{1}{v} \frac{\partial}{\partial t} + j (\nabla \wedge \cdot)$$

and applying the modified Maxwell's equations mentioned above, we obtain the final *direct matrix generator* Γ of the matrix wavelets

$$\begin{bmatrix} \Psi_n^{E'} \\ \Psi_n^{H'} \end{bmatrix}$$

as

$$\alpha \begin{bmatrix} \Psi_{n+1}^{E'} \\ \Psi_{n+1}^{H'} \end{bmatrix} = \underbrace{\begin{bmatrix} \frac{1}{v} \frac{\partial}{\partial t} & -j\mu \left(\frac{\partial}{\partial t} - \frac{\sigma}{2\varepsilon} \right) \\ j\varepsilon \left(\frac{\partial}{\partial t} - \frac{\sigma}{2\varepsilon} \right) & \frac{1}{v} \frac{\partial}{\partial t} \end{bmatrix}}_{\Gamma} \begin{bmatrix} \Psi_n^{E'} \\ \Psi_n^{H'} \end{bmatrix}$$

A *reverse matrix generator* can be found in the same way by dealing with the recursive equations $\alpha \Psi_{n-1}^{E'} = G' \Psi_n^{E'}$ and $\alpha \Psi_{n-1}^{H'} = G' \Psi_n^{H'}$, with

$$G' = \frac{1}{v} \frac{\partial}{\partial t} - j(\nabla \wedge \cdot)$$

but it is useless for computing the wavelet family from its seed.

9.2.4. *Spherical seed and wavelets having a zero divergence*

The previous development, focused on the commutative factorization of a second-order partial differential operator, has not accounted for the necessary zero divergence of the electromagnetic field. Yet, the previous wavelet generating equation easily shows the zero divergence of the wavelet seed corresponding to $n = 0$ involves the same for any $n > 0$.

The required zero divergence is obtained by means of the electrical and magnetic fields being the "curl" vectors (that are the result of the operator $(\nabla \wedge \cdot)$ on vectors) of vector potentials defined in Chapter 1. These vector potentials meet the lossy propagation equation according to the Lorentz gauge (see Chapter 1).

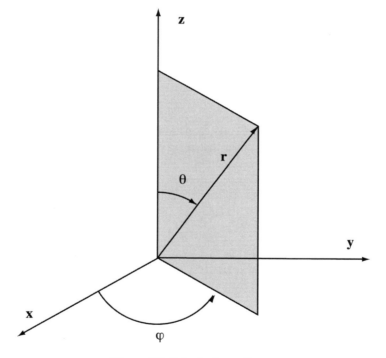

Figure 9.2. *Spherical coordinates*

It is well known in the classical theory of Green's functions for time-varying electromagnetic fields that any solution of the even lossy propagation equation is the *convolution* product in the space domain of a "Green's function" defined as a wave radiating outwards from a point source with a given spatial distribution of point sources.

The Green's function tied to a point source necessarily has a spherical symmetry. Therefore spherical coordinates (Figure 9.2) are involved in solving the lossy propagation equation.

The spherical symmetry reduces the classical expression of the Laplace operator

$$\nabla^2 = \frac{1}{r}\frac{\partial^2}{\partial r^2}(r\cdot) + \frac{1}{r^2 \sin\theta}\frac{\partial}{\partial \theta}\left(\sin\theta \frac{\partial}{\partial \theta}\right) + \frac{1}{r^2 \sin\theta}\frac{\partial^2}{\partial \varphi^2}$$

to its first term so that the lossy propagation equation becomes

$$\left[v^2 \frac{\partial^2}{\partial r^2} - \frac{\partial^2}{\partial t^2} - 2\alpha \frac{\partial}{\partial t}\right](r\vec{A}) = 0$$

This is a telegrapher's equation the solution of which corresponding to a given time-varying source has been represented by a wavelet expansion. Dividing it by r gives new 3D wavelets having spherical symmetry corresponding to a given point source:

$$\vec{A}(r,t) = \frac{1}{r}\sum_{n=0}^{\infty} \vec{a}_n \alpha^{-n} \left(\frac{t - \frac{r}{v}}{t + \frac{r}{v}}\right)^{n/2} I_n\left(\alpha\sqrt{t^2 - \frac{r^2}{v^2}}\right) e^{-\alpha t}$$

In our application, $\alpha = \sigma/\varepsilon$, so multiplying the parts of the last relation by $e^{\alpha t}$ leads directly to the wavelets $\Psi_n^{E'}$ and $\Psi_n^{H'}$ first with $n = 0$ then $n > 0$ by means of

$$\alpha \begin{bmatrix} \Psi_{n+1}^{E'} \\ \Psi_{n+1}^{H'} \end{bmatrix} = [\Gamma] \begin{bmatrix} \Psi_n^{E'} \\ \Psi_n^{H'} \end{bmatrix}$$

9.2.5. *Modeling electromagnetic perturbations in lossy media*

The previous wavelet expansion can be used as a perturbing source at the input terminal of a waveguide which has been modeled by an equivalent transmission line generally loaded by distributed vector potential sources at the end of the chapter dealing with straight homogenous media. This can be the case for any electrical connection traveled by very high-speed signals.

Another application situation is that of a perturbation created on a lossless plane conductor by a pulsed electromagnetic source far away from it. In this case, the method of electromagnetic images can be applied according to Chapter 1.

Considering that the pulsed source creates the vector potential $\vec{A}(r,t)$ where $r = \sqrt{(x-x_0)^2 + y^2 + z^2}$ according to Figure 9.2 at the point (x, y, z), its image through the plane creates the vector potential $\vec{A}(r,t')$ where $r' = \sqrt{(x+x_0)^2 + y^2 + z^2}$ at the same point. On the even plane $x = 0$, the pulsed source creates a current having a density given by

$$\vec{j} = 2\vec{n} \wedge (\vec{\tilde{\nabla}} \wedge \vec{A})$$

where $\vec{n}$ is the unitary vector normal to the plane at the point $(0, y, z)$.

9.2.6. *Guided propagation in interconnection structures*

Interconnections inside electronic substrates (integrated circuits, printed circuit boards, etc.) have local structure such as "microstrip" which is able to be modeled by means of cylindrical wavelets. These are developed in Appendix E, and their application to guided propagation is highlighted in Appendix F.

9.3. Conclusion

By means of the introduction of propagation wavelets, a matrix equation related to any complex network made up of lossy conductors and dielectrics can be written directly in the time domain without the need of frequency domain, convolution product, or gain time discretization. This allows us to compute the amplitude of many-times-reflected waveforms in one "shot" even with the existence of lumped elements between the distributed parameter segments of lines. The wavelet method has been extended to solve the 3D problems of electromagnetic compatibility in lossy media.

Appendix A

Physical Data

This appendix concerns some numerical values of the electromagnetic constants encountered in Chapter 1: dielectric permittivity, magnetic permittivity and electrical conductivity.

To begin with, in respect of the use of international units, we are obliged to recall the "dimensions equations" related to any physical features, and therefore to the electromagnetic parameters. Let us recall that these "dimensions equations" are monomial formulae of any physical feature expressed in terms of a few considered as *reference features*. The following are these reference features:

– the *length* denoted as L, given in meters;
– the *mass* denoted as M, given in kilograms;
– the *time* denoted as T, given in seconds;
– the *intensity of electrical current* denoted as I, given in amperes.

Then, we can define the units corresponding to the electromagnetic features:

– the *electrical field*, $LMT^{-3}I^{-1}$, given in volts per meter;
– the *magnetic field*, $L^{-1}I$, given in amperes per meter;
– the *electrical displacement*, $L^{-2}TI$;
– the *magnetic induction*, $MT^{-2}I^{-1}$, given in teslas;
– the *scalar potential*, $L^2MT^{-3}I^{-1}$, given in volts;
– the vector potential and the *electrical current density*, $L^{-2}I$, given in amperes per square meter.

So, the units of the electromagnetic constants can be considered:

– the *dielectric permittivity*, defined as the electrical displacement divided by the electrical field, $L^{-3}M^{-1}T^4I^2$;

– the *magnetic permeability*, defined as the magnetic induction divided by the magnetic field, $LMT^{-2}I^{-2}$;

– the *electrical conductivity*, defined as the electrical current density divided by the electrical field, $L^{-3}M^{-1}T^3I^2$.

Now, it is possible to obtain data about features associated with the electromagnetic properties of various materials with regard to the units mentioned above.

Let us start with dielectric permittivity. That of a vacuum, denoted as ε_0, is equal to $1/(36\pi \times 10^9)$, to which any other dielectric permittivity is related by means of a multiplicative coefficient called the "relative dielectric permittivity" of each material, denoted as ε_r. Some values of ε_r are as follows:

– vacuum = 1;
– organic materials: polyimides = 3.5; epoxy = 4.2;
– for any mix of organic materials having close values of ε_n [LAN4]:

$$\varepsilon_{mix}^{1/3} = \frac{1}{N}\sum_{n=1}^{n=N}\varepsilon_n^{1/3}$$

Note that these materials show a non-zero electrical conductivity associated with water diffusion inside them. This non-zero electrical conductivity comes into the harmonic wave equations as developed in Chapter 2 since the frequency is higher than 1 MHz. This can be simulated by modifying the relative dielectric permittivity and omitting the term corresponding to the finite electrical conductivity of the material. In so doing, an imaginary further term is added to the real relative electrical permittivity. The modulus of this imaginary term is equal to the product of the so-called loss angle which depends on the material and the low-frequency (typically less than 1 kHz) harmonic value of the relative electrical permittivity. This loss angle has a range of between 1/100 and 1/10,000 of the relative electrical permittivity. Conversely, knowing this loss angle leads to an easy estimation of the leakage electrical conductivity of the material and therefore the terms of the matrix [G] introduced in Chapters 4, 6, 7, 8 and 9.

Some further values of ε_r are as follows:

– ceramics: alumina = 9; titanium oxide = 150; titanium barium oxide = 1,500; dielectric capacitors = from 30 to 100,000;

– intrinsic semiconductors: gallium arsenide = 10.9; silicon = 11.9; indium phosphide = 12.4.

The magnetic permittivity of a vacuum denoted as μ_0 is equal to $4\pi \times 10^{-7}$, to which any other magnetic permittivity is related by means of the "relative magnetic permittivity" denoted as μ_r. This is equal to 1 for the materials mentioned above. However, this relative magnetic permittivity becomes greater than 1, up to several thousands in the case of the so-called "magnetic" materials, but strongly decreases when the frequency of the magnetic field increases.

We conclude with a discussion of electrical conductivity, some data for which are given below, with values given in millions of units:

– silver = 62.5;
– copper = 58.8;
– gold = 42.5;
– aluminum = 37.7;
– molybdenum = 19.2;
– tungsten = 17.7;
– nickel = 14.6;
– palladium = 9.1;
– chromium = 7.7.

For a conductive powder made up of metal particles having an electrical conductivity σ_c within an insulating material with a volume ratio η, the resulting electrical conductivity is given by

$$\sigma_{powder} = \frac{1}{2}(3\eta - 1)\sigma_c$$

which shows that the electrical conductivity is zero when the volume ratio is less than 1/3.

Inside n-doped (respectively p-doped) semiconductor media the electrical conductivity is $\sigma_n = 1.6 \times 10^{-20} n$ (respectively $\sigma_p = 4.8 \times 10^{-21} p$), n and p being the concentration of impurities in atoms per cubic meter [VAP2]. Its value is set between 10^{-6} and 1, then $\sigma/2\varepsilon$ is between 10^5 and 10^{11} for semiconductor layers.

The value reaches 10^{19} for bulk metals. The value of $a = \mu\sigma^2/4\varepsilon$ goes from 10^{-7} (weakly doped semiconductor media) to 10^{19} (metals).

Appendix B

Technological Data

	Printed circuits	Thin films	Integrated circuits
Conductor materials	Copper, tin, nickel, gold	Copper, nickel, gold	Aluminum, titanium, gold
Conductor width (μm)	200 to 50	50 to 5	5 to 0.2
Conductor thickness (μm)	70 to 5	5 to 1	1 to 0.1
Dielectric materials	Epoxy, polyimides	Polyimides, alumina	Silicon, SiO_2, AsGa
Conductor spacing (μm)	200 to 50	50 to 5	5 to 0.2
Dielectric thickness between layers (μm)	100 to 25	25 to 2	2.5 to 0.05
Number of layers	2 to 24	2 to 8	2 to 4

Appendix C

Lineic Capacitors

This appendix deals with lineic capacitor computations as introduced in Chapter 4 with the Laplace equation:

$$\underbrace{\left(\frac{\partial^2}{\partial x^2} + \frac{\partial^2}{\partial y^2}\right)}_{\nabla_s^2} u = 0$$

where the "two-dimensional" (2D) elliptic Laplacian operator is highlighted. We go on to model the quasi-TEM electromagnetic field by means of an analytical approach suited for the particular geometry of the straight conductor cross-sections met in standard interconnections used in electronics.

Indeed, the cross-sections of the straight conductors met in classical technological structures have simple geometrical shapes such that they can be represented analytically in a system of *orthogonal curvilinear coordinates*.

Let $\{q_1, q_2\}$ be the set of orthogonal curvilinear coordinates formed on the boundary of the cross-sections of one or several conductors having similar shapes in terms of differential topology. The square of the elementary length is written as $ds^2 = e_1^2 dq_1^2 + e_2^2 dq_2^2$. Then, the "2D" Laplacian becomes

$$\nabla_s^2 = \frac{1}{e_1 e_2}\left[\frac{\partial}{\partial q_1}\left(\frac{e_2}{e_1}\frac{\partial}{\partial q_1}\right) + \frac{\partial}{\partial q_2}\left(\frac{e_1}{e_2}\frac{\partial}{\partial q_2}\right)\right]$$

Inside the technological structures of interconnections, it is possible to choose, in this case, that the Laplace equation remains in its Cartesian coordinate form:

$$\left[\frac{\partial^2}{\partial q_1^2} + \frac{\partial^2}{\partial q_2^2}\right]u = 0$$

After the Laplace equation has been solved in the new coordinate system suited for the conductor shape, u designating the scalar potential which has to be constant on the surface ("Dirichlet conditions"), the electrical field coordinates are given by

$$\frac{1}{e_1}\frac{\partial u}{\partial q_1}, \quad \frac{1}{e_2}\frac{\partial u}{\partial q_2}$$

So, the lineic density of electrical charges is equal to the derivative $\partial u/\partial n$ across the conductor surface. Integrating it along the boundary path of the cross-section of the conductor and multiplying it by the dielectric permittivity yields

$$\varepsilon \oint_{\text{section}} \frac{\partial u}{\partial n}\, dl$$

which is the total electrical charge. Setting a zero scalar potential on every conductor except one, the capacitor of the non-zero scalar potential conductor is equal to the inverse of its scalar potential when its has a unitary charge.

In particular, choosing one of the coordinates as constant on the conductor surface, u is linearly dependent on the other coordinate and the computation of the mutual capacitor between this conductor and its neighbors becomes the obvious one of a plane capacitor; this is the main advantage of the orthogonal curvilinear coordinates applied to interconnection modeling.

In the more general case of non-constant coordinates on the conductor boundary, using the very classical variable separation method leads to expansions in products of eigenfunctions each one depending on a single variable and having to be constant on the conductors boundary ("Dirichlet condition"):

$$u = \sum_{m,n} a_{m,n} f_m(q_1) g_n(q_2)$$

In what follows, we present the main interesting orthogonal curvilinear coordinates suited for interconnection modeling.

Parabolic coordinates: α and β, related to the Cartesian coordinates by

$$x = k\alpha\beta, \quad y = \frac{k}{2}(\beta^2 - \alpha^2)$$

They are suited for processing the *electromagnetic field in the vicinity of a half-plane edge*: $x = 0, y \geq 0$, so $\alpha = 0$.

Elliptic coordinates: ξ and φ, related to the Cartesian coordinates by

$$x = a\operatorname{ch}\xi \cos\varphi, \quad y = a\operatorname{sh}\xi \sin\varphi$$

They are suited for processing the *electromagnetic field in the vicinity of one or several parallel straight thin conductors or again of a straight opening in a conductive plane*.

Indeed, $\varphi = 0$ represents a straight opening of width a into a conductive plane, $\xi = 0$ represents a thin conductor having width a. Any non-zero ξ represents an elliptic cross-section conductor (φ varying from 0 to 2π) or again a semi-elliptic one (φ varying from 0 to π), this case corresponding to a metal coating on any interconnection substrate.

Let w be the width of a metal coating having thickness t. Then the orthogonal curvilinear coordinate system is defined by

$$\operatorname{th}\xi = \frac{t}{w} \quad a = \frac{1}{2}\sqrt{w^2 - t^2}$$

This theoretical modeling can be improved in order to account for what is called an under-etch on the edges of the metal coating by cutting the semi-elliptic cross-section symmetrically by two hyperbolic arcs corresponding to a non-zero value of φ. Let w be the width of the upper side of the metal coating, g being the width at its bottom side and t the greatest thickness. We obtain

$$\operatorname{ch}\xi = \frac{w}{g} \qquad \cos\varphi = \frac{w}{t}\operatorname{th}\xi$$

$$a = \frac{1}{2}\sqrt{\left(\frac{w}{\cos\varphi}\right)^2 - t^2}$$

Any set of thin conductors which are parallel and non-overlapping can be theoretically depicted by the intersection of several elliptic cylinders corresponding to different values of ξ with the two half-planes around the opening corresponding to $\varphi = 0$. In the case of equal spacing and width, the values of ξ are solutions of $\operatorname{ch}\xi = 2n+1$, $n < N$ being the number of conductors on one side of the central one.

Biaxial coordinates: ξ and φ, related to the Cartesian coordinates by

$$x = a\frac{\operatorname{sh}\xi}{\operatorname{ch}\xi + \cos\varphi}, \qquad y = a\frac{\sin\varphi}{\operatorname{ch}\xi + \cos\varphi}.$$

They are suited for processing the *electromagnetic field around two circular wires or one wire above a conductive plane*. Each couple of wires corresponds to a non-zero value of ξ while the middle plane parallel to the wires having the same ζ corresponding to $\xi = 0$. The greater is ξ the thinner is the wire diameter.

It is possible to transform the biaxial coordinates by dilatation along either x or y so that they approach the elliptic shape of metal coating of conductors on any interconnection substrate. Dilating x deals with two metal lines set on the same interconnection layer.

Let w be the similar width of the conductors, t and s being their thickness and spacing, respectively, defining the dilation coefficient k:

$$k = \frac{w}{t} \qquad \operatorname{sh}^2\frac{\xi}{2} = \frac{s}{2w} \qquad a = \frac{1}{2}w\operatorname{sh}\xi$$

Dilating y deals with two superposed elliptic conductors belonging to different interconnection layers or a single elliptic conductor parallel to a metal plane corresponding to $\xi = 0$. The previous formulae are still valid since k is the y dilatation and $h = w/2$ is the distance of each conductor to the metal plane.

In the previous two curvilinear coordinate systems, we have to get the solutions of the Laplace equation which has the same form in these systems as in the Cartesian coordinates. This means

$$\frac{\partial^2 u}{\partial \xi^2} + \frac{\partial^2 u}{\partial \varphi^2} = 0$$

Accounting for the periodicity versus φ:

$$u = \sum_{m=-\infty}^{m=+\infty} \left[a_m e^{m(\xi + j\varphi)} + \overline{a_m} e^{m(\xi - j\varphi)} \right]$$

where $\overline{a_m}$ is the complex conjugate of a_m.

u not depending on ξ along any curve corresponding to $\varphi = \varphi_0$ yields that the coefficient of any $e^{m\xi}$ is zero:

$$a_m e^{jm\varphi_0} + \overline{a_m} e^{-jm\varphi_0} = 0$$

while u being constant on any curve $\xi = \xi_0$ set between φ_1 and φ_2 requires that the coefficient of any $e^{jm\varphi}$ is equal to the corresponding one of the Fourier expansion of the well-known "gate" function defined as constant (arbitrary value u_0) between φ_1 and φ_2:

$$a_m e^{m\xi_0} + \overline{a_{-m}} e^{-m\xi_0} = u_0 \frac{\sin[(m/2)(\varphi_2 - \varphi_1)]}{(m/2)(\varphi_2 - \varphi_1)} e^{-j(m/2)(\varphi_1 + \varphi_2)}$$

To the last two equations have been added those obtained by changing m into $-m$. Thus, we obtain a linear system made of four equations giving the real and imaginary parts of a_m and a_{-m} that is related to a given shape of the conductor cross-section set at different potentials. Of course, these coefficients are multiplied by any unknown factor like u_0.

We can deal with the curvilinear coordinates of the quasi-TEM electrical field:

$$E_\xi = -\frac{1}{h(\xi, \varphi)} \frac{\partial u}{\partial \xi} \qquad E_\varphi = -\frac{1}{h(\xi, \varphi)} \frac{\partial u}{\partial \varphi}$$

with

$$h(\xi, \varphi) = a\sqrt{\operatorname{ch}^2 \xi - \cos^2 \varphi}$$

for elliptic coordinates and

$$h(\xi,\varphi) = \frac{a}{\operatorname{ch}\xi + \cos\varphi}$$

for biaxial coordinates (transformed or not).

Thus, the lineic charge density over the cross-section of any conductor can be computed as

$$\sigma = \varepsilon E_\varphi \quad \text{onto any arc corresponding to a constant } \xi$$

and

$$\sigma = \varepsilon E_\xi \quad \text{onto any arc corresponding to a constant } \varphi$$

The total electrical charge is given by

$$Q = \varepsilon \int_{\text{arcs } \xi \text{ constant}} \sigma(\xi,\varphi) h(\xi,\varphi) \, d\varphi + \varepsilon \int_{\text{arcs } \varphi \text{ constant}} \sigma(\xi,\varphi) h(\xi,\varphi) \, d\xi$$

The lineic capacitor becomes

$$C = \frac{Q}{u} = \frac{Q}{u_0}$$

The factor u_0 similarly multiplies the scalar potential, the electrical field coordinates, and the electrical charge density; therefore the value of the lineic capacitor is not changed if $u_0 = 1$ is chosen which yields the required result:

$$C = \varepsilon \int_{\text{arcs } \xi \text{ constant}} \frac{\partial u}{\partial \xi} \, d\varphi + \varepsilon \int_{\text{arcs } \varphi \text{ constant}} \frac{\partial u}{\partial \varphi} \, d\xi$$

Appendix D

Modified Relaxation Method

In Chapters 5–9, we solved propagation equations written either in the time or frequency domain and converted them into a matrix equation as follows:

$$[A][x] = [y]$$

Let us recall briefly the general purpose relaxation method aimed at solving the above equation. It is known that this method is a recursive one, so is based on running a set of recurrence equations that have to be stopped by means of any criterion of convergence. According to the theory of this method, the convergence can be met rigorously only if the square matrix of the equation is Hermitian (its elements set symmetrically with regards to the diagonal are conjugate complex numbers) and positive definite (then the matrix is associated with a positive quadratic form, so $[x]^*[A][x] \geq 0$, where $[x]^*$ is both the transposed and conjugate vector of $[x]$).

This is not the case for the matrix $[A]$; however, if we multiply it on its left side by its transposed and conjugate matrix denoted as $[A]^*$, the result becomes hermitian and positive definite. Performing this left-side multiplication on the parts of the matrix equation yields

$$\underbrace{[A]^*[A]}_{[B]}[x] = \underbrace{[A]^*[y]}_{[z]}$$

Now, we can go on to the rigorous application of the relaxation method. Let us choose a parameter ω so that the convergence process is accelerated, a computation of the n coordinates x_i of the vector $[x]$ runs recursively depending on the coordinates z_i of the vector $[z]$ and of the elements b_{ij} of matrix $[B]$ defined above. Then, at the $k+1$ step of this recursive computation, we obtain

$$x_i^{k+1} = \underbrace{\omega\left\{\underbrace{\frac{1}{b_{ii}}\left(z_i - \sum_{j=1}^{j=i-1} b_{ij} x_j^{k+1} - \sum_{j=i+1}^{j=n} b_{ij} x_j^k\right)}_{\text{Gauss-Seidel algorithm}}\right\} + (1-\omega) x_i^k}_{\text{relaxation method}}$$

The convergence criterion becomes rigorously

$$0 < \omega < 2$$

Programming the above computations involves the advantages of gathering the non-zero elements of the matrix to be inverted close to the main diagonal by means of ordering of all the segments of the nets developed in section 2.3.

Appendix E

Cylindrical Wavelets

This appendix deals with the wavelet applications of Chapter 9, mainly the 3D applications. Let the propagation equation be

$$\left(\underbrace{\frac{\partial^2}{\partial x^2} + \frac{\partial^2}{\partial y^2}}_{\nabla_s^2} + \underbrace{\frac{\partial^2}{\partial z^2} - \frac{2\alpha}{v^2}\frac{\partial}{\partial t} - \frac{1}{v^2}\frac{\partial^2}{\partial t^2}}_{T} \right) u = 0$$

where the 2D elliptic (or Laplacian) operator (∇_s^2) and the Telegrapher operator (T) are highlighted, as already noted in Chapter 4. This is applicable in the vicinity of straight conductors only.

According to Appendix C, eigenfunctions tied to the eigenvalue λ of the 2D Laplacian operator have been found in any curvilinear coordinate system so that this operator can be written as $\partial^2/\partial\xi\partial\bar{\xi}$. Applying the results already obtained in Appendix C, we are led to a solution made up of functions where the four variables $\xi, \bar{\xi}, z, t$ are shared into two couples ($\xi, \bar{\xi}$) and (z, t), respectively:

$$u(\xi,\bar{\xi},z,t,\lambda) = \sum_{n=0}^{\infty} Z_n(z,t) \underbrace{\left(\frac{\bar{\xi}}{\xi}\right)^{n/2} I_n(2|\xi|\sqrt{\lambda})}_{\Psi_n(\xi,\bar{\xi},\lambda)} = \sum_{n=0}^{\infty} \Psi_n(\xi,\bar{\xi},\lambda) Z_n(z,t)$$

We notice that this solution defines a vector space structure for the set of functions depending on ξ and $\bar{\xi}$ and spanned by $\{\Psi_n\}$. Setting this solution into the propagation equation $(\nabla_s^2 + T)u = 0$, we obtain

$$(\nabla_s^2 + T)\sum_{n=0}^{\infty} \Psi_n Z_n = \sum_{n=0}^{\infty} \left[(\nabla_s^2 \Psi_n) Z_n + \Psi_n (TZ_n) \right] = \sum_{n=0}^{\infty} \left[(\lambda \Psi_n) Z_n + \Psi_n (TZ_n) \right]$$

$$= \sum_{n=0}^{\infty} \Psi_n [\lambda Z_n + TZ_n] = 0$$

The vector space structure induced by $\{\Psi_n\}$ requires that the functions $Z_n(z,t)$ have to satisfy $(T + \lambda)Z_n = 0$, already solved by means of the wavelet expansion in Chapter 8 in the particular case of $\lambda = 0$.

Then, we obtain

$$Z_n(z,t) = \sum_{m=0}^{\infty} a_{m,n} \Phi_m(z,t)$$

Indeed, $(T + \lambda)u = 0$, so explicitly

$$\left(\frac{\partial^2}{\partial z^2} - \frac{2\alpha}{v^2} \frac{\partial}{\partial t} - \frac{1}{v^2} \frac{\partial^2}{\partial t^2} + \lambda \right) u = 0$$

Let us come back to the factorization of operators developed in Chapter 8 by means of the transformation $u = u_1 e^{-\alpha t}$. This gives

$$\left(\frac{\partial^2}{\partial t^2} - v^2 \frac{\partial^2}{\partial z^2} + \lambda \right) u_1 = (\alpha^2 + \lambda v^2) u_1 = \alpha_1^2 u_1$$

and the wavelets

$$\Phi_m(z,t) = \alpha_1^m \left(\frac{t - \frac{z}{v}}{t + \frac{z}{v}} \right)^{m/2} I_m \left(\alpha_1 \sqrt{1 - \frac{z^2}{t^2}} \right) e^{-\alpha t}$$

Coming back to u, we obtain the required general solution as a double expansion as follows:

$$u(\xi,\overline{\xi},z,t,\lambda) = \sum_{m,n} a_{m,n} \underbrace{\Psi_n(\xi,\overline{\xi},\lambda)\Phi_m(z,t)}_{\text{cylindrical wavelets}}$$

which defines what we propose calling "cylindrical wavelets".

Appendix F

Wavelets and Elliptic Operators

This appendix completes Chapter 9 related to wavelet applications, in particular the 3D applications. In order to solve the propagation equation, let the equations of eigenfunctions tied to the eigenvalue λ be

$$\underbrace{\left(\frac{\partial^2}{\partial x^2} + \frac{\partial^2}{\partial y^2}\right)}_{\nabla_s^2} u = \lambda u$$

where we have highlighted the 2D elliptic operator (or "two-dimensional" (2D) Laplacian). This assumes, obviously, that we are interested in straight conductors having any cross-section in homogenous dielectric media, as is the case for *electronic board interconnections*.

Then, we apply the wavelet generating process developed in Chapter 8 in the space domain related to the "2D" Laplacian. According to the main classical interconnection structures used in electronic boards, the cross-section of straight conductors has a simple geometrical shape such that it is possible to depict it by means of *orthogonal curvilinear coordinates*.

Let $\{q_1, q_2, z\}$ be the orthogonal curvilinear coordinates along axes formed on the cross-section of one or several straight conductors in the case of these having a similar shape in terms of differential topology. The square of the elementary length in any cross-section is written as $ds^2 = e_1^2 dq_1^2 + e_2^2 dq_2^2$. Then the "2D" Laplacian becomes

$$\nabla_s^2 = \frac{1}{e_1 e_2} \left[\frac{\partial}{\partial q_1} \left(\frac{e_2}{e_1} \frac{\partial}{\partial q_1} \right) + \frac{\partial}{\partial q_2} \left(\frac{e_1}{e_2} \frac{\partial}{\partial q_2} \right) \right]$$

This operator can be shared into factors in terms of generating wavelets discussed in Chapter 8, and it is possible to write $e_1 = e_2 = |h(q_1, q_2)|$, with

$$|h(q_1, q_2)|^2 = h\underbrace{\left(q_1 + jq_2 \right)}_{\kappa} h\underbrace{\left(q_1 - jq_2 \right)}_{\bar{\kappa}}$$

Then the Laplacian is shared which defines two new complex variables, χ and its conjugate:

$$\nabla_s^2 = \frac{1}{|h(q_1,q_2)|^2} \left(\frac{\partial^2}{\partial q_1^2} + \frac{\partial^2}{\partial q_2^2} \right) = \frac{1}{|h(q_1,q_2)|^2} \left(\frac{\partial}{\partial q_1} + j\frac{\partial}{\partial q_2} \right)\left(\frac{\partial}{\partial q_1} - j\frac{\partial}{\partial q_2} \right)$$
$$\underbrace{}_{2\frac{\partial}{\partial \bar{\kappa}}} \underbrace{}_{2\frac{\partial}{\partial \kappa}}$$

$$= \underbrace{\frac{2}{h(\kappa)} \frac{\partial}{\partial (\kappa)}}_{\frac{\partial}{\partial \chi}} \left[\underbrace{\frac{2}{h(\bar{\kappa})} \frac{\partial}{\partial (\bar{\kappa})}}_{\frac{\partial}{\partial \bar{\chi}}} \right] = \frac{\partial^2}{\partial \chi \partial \bar{\chi}}$$

from which we find

$$\chi = \frac{1}{2} \int h(\kappa) \, d\kappa \quad \text{and} \quad \bar{\chi} = \frac{1}{2} \int h(\bar{\kappa}) \, d\bar{\kappa}$$

The modulus is denoted as $|\chi| = \sqrt{\chi \bar{\chi}}$.

The eigenfunctions of the operator $\partial^2 / \partial \chi \partial \bar{\chi}$ tied to the eigenvalue λ have been solved in Chapter 8 by means of the Picard method, thus leading directly to the following wavelet expansion:

$$u = \sum_{n=0}^{\infty} a_n \left(\frac{\bar{\chi}}{\chi} \right)^{n/2} I_n(2|\chi|\sqrt{\lambda}) = \sum_{n=0}^{\infty} a_n I_n(2|\chi|\sqrt{\lambda}) e^{-jn(\arg \chi)}$$

The coefficients of this expansion are dependent on the boundary conditions on the boundary of the cross-section represented in orthogonal curvilinear coordinates. Moreover, it has been shown in Chapter 4 that the eigenvalues of $(-\nabla_s^2)$ are positive so we can set $\lambda = -\gamma^2$ and, applying $I_n(2|\chi|\sqrt{\lambda}) = j^n J_n(2\gamma|\chi|)$, we obtain

$$u = \sum_{n=0}^{\infty} a_n J_n(2\gamma|\chi|) e^{-jn[\arg(\chi) - \pi/2]}$$

Such is the case for direct wavelets. Changing n to $(-n)$ gives reverse wavelets.

$n \in Z, Z$ being the set of relative integers, we get a superposition of direct and reverse wavelets.

The following are some orthogonal curvilinear coordinates for which the method works well.

Cartesian coordinates: x and y. We obtain

$$|h|^2 = 1, \quad \chi = \frac{\kappa}{2} = \frac{x + jy}{2}, \quad \bar{\chi} = \frac{x - jy}{2}, \quad \arg(\chi) = \tan^{-1}\frac{y}{x} = \theta,$$

$$|\chi| = \frac{1}{2}\sqrt{x^2 + y^2} = \frac{r}{2}$$

This shows an original way to represent the solution u as an already well-known expansion of any wave in polar coordinates in Bessel functions. This well-known expansion is classically found by means of the separation of variables r and θ in the eigenfunction equation. In any case, the result is

$$u = \sum_{n=0}^{\infty} a_n e^{-jn(\theta - \pi/2)} J_n(\gamma r)$$

which can be applied to propagation in a *circular waveguide* as an *optical fiber* for instance.

Parabolic coordinates: α and β, related to Cartesian coordinates by

$$x = k\alpha\beta, \quad y = \frac{k}{2}(\beta^2 - \alpha^2)$$

These are suited for *processing an electromagnetic field in the vicinity of the edge of a half-plane* $x = 0, y \geq 0$, which corresponds to $\alpha = 0$. For any non-zero value of α and inside any range of β, we obtain a modeling of a *cylindrical parabolic antenna*.

Separating the variables α and β in the eigenfunction equation leads to classical solutions expanded in series of *Weber–Hermite functions*. We move to another way:

$$|h|^2 = k^2(\alpha^2 + \beta^2), \quad h = k(\alpha + j\beta) = k\kappa$$

so

$$\chi = k \frac{\kappa^2}{4}$$

then

$$u = \sum_{n=0}^{\infty} a_n J_n\left(\frac{\gamma k}{2}(\alpha^2 + \beta^2)\right) e^{-jn[2\tan^{-1}(\beta/\alpha) - \pi/2]}$$

Elliptic coordinates: ξ and φ, related to Cartesian coordinates by

$$x = a\operatorname{ch}\xi \cos\varphi, \quad y = a\operatorname{sh}\xi \sin\varphi$$

These are suited for processing an *electromagnetic field in the vicinity of a straight opening in a plane or around a thin straight conductor*. Condition $\varphi = 0$ gives an opening having width a; $\xi = 0$ gives a thin conductor having width a. Any non-zero ξ gives an *elliptic waveguide (or optical fiber)*.

Separating the variables ξ and φ in the eigenfunction equation leads to classical solutions expanded in series of *Mathieu functions* that are very difficult to handle. We move to another way:

$$|h|^2 = a^2(\operatorname{ch}^2\xi - \cos^2\varphi) = a^2 \operatorname{sh}\underbrace{(\xi + j\varphi)}_{\kappa} \operatorname{sh}(\xi - j\varphi)$$

then

$$h(\kappa) = a\operatorname{sh}\kappa \quad \text{and} \quad \chi = \frac{a}{2}\operatorname{ch}\kappa = \frac{a}{2}(\operatorname{ch}\xi \cos\varphi + j\operatorname{sh}\xi \sin\varphi)$$

$$u = \sum_{n=0}^{\infty} a_n J_n \left(\gamma a \sqrt{\text{ch}^2 \xi \cos^2 \varphi + \text{sh}^2 \xi \sin^2 \varphi} \right) e^{-jn[\tan^{-1}(\text{th}\xi \tan \varphi) - \pi/2]}$$

When $\xi = 0$ we obtain the expansion

$$u = \sum_{n=0}^{\infty} a_n j^n J_n (\gamma a \cos \varphi)$$

which can be considered as a boundary condition. Known complex relationships between a_n give a classical representation of the Mathieu functions.

If $a_{n \in Z} = 1$, we obtain $u = e^{j\gamma a \text{ch}\xi \cos \varphi}$ which represents a plane wave expanded into *direct and reverse wavelets* traveling along the x axis the influence of which (induced current) on any thin straight conductor can be rigorously computed.

Biaxial coordinates: ξ and φ, related to Cartesian coordinates by

$$x = a \frac{\text{sh}\xi}{\text{ch}\xi + \cos \varphi}, \quad y = a \frac{\sin \varphi}{\text{ch}\xi + \cos \varphi}$$

These are suited for processing an *electromagnetic field in the vicinity of a circular wire above a plane or of two wires having different diameters*. Any wire corresponds to a non-zero value of ξ while the plane parallel to this wire can be represented by $\xi = 0$. The greater the value of ξ, the thinner the wire.

We can write

$$|h|^2 = \frac{a^2}{(\text{ch}\xi + \cos \varphi)^2} = \frac{a^2}{4 \text{ch}^2 \frac{\xi + j\varphi}{2} \text{ch}^2 \frac{\xi - j\varphi}{2}}$$

then

$$h(\kappa) = \frac{a}{2\text{ch}^2 \frac{\kappa}{2}} \quad \text{and} \quad \chi = \frac{a}{2} \text{th} \frac{\kappa}{2} = \frac{a}{2} \frac{\text{sh}\xi + j \sin \varphi}{\text{ch}\xi + \cos \varphi}$$

$$u = \sum_{n=0}^{\infty} a_n J_n \left(\gamma a \frac{\sqrt{\text{sh}^2 \xi + \sin^2 \varphi}}{\text{ch}\xi + \cos \varphi} \right) e^{-jn[\tan^{-1}(\sin \varphi / \text{sh}\xi) - \pi/2]}$$

If $a_{n \in Z} = 1$, we get a plane wave expanded into *direct and reverse wavelets* traveling along the *y* axis on two wires.

References

[AHM] A. AHMADOUCHE, J. CHILO, "Optimum computation of capacitance coefficients of multilevel interconnecting lines for advanced package", *IEEE Transactions on Components, Hybrids, and Manufacturing Technology*, vol. 12, 124-129, 1989.

[ANG1] A. ANGOT, *Compléments de Mathématiques à l'usage des Ingénieurs de l'Electrotechnique et des Télécommunications*, Chapter III, p. 109-114, Editions de la Revue d'Optique, Paris, 1961.

[ANG2] A. ANGOT, *Compléments de Mathématiques à l'usage des Ingénieurs de l'Electrotechnique et des Télécommunications*, p. 384-385, Editions de la Revue d'Optique, Paris, 1961.

[BAN] C. BANDLE, *Isoperimetric Inequalities and Application*, p. 128, Pitman, London, 1980.

[BOS] A. BOSSAVIT, "On the finite elements for the electricity equations", *The Mathematics of Finite Elements and Applications IV*, pp.85-91, J.R. Whiteman, Academic Press, 1982.

[BOU] M. BOUIX, *Les discontinuités du rayonnement électromagnétique*, Chapter 1, p. 5-7, Chapter 2, p. 35-38, Editions Dunod, 1965.

[BRE1] H. BREZIS, *Analyse fonctionnelle*, Chapter V, p. 78, Editions Masson, 1992.

[BRE2] H. BREZIS, *Analyse fonctionnelle*, Chapter IX, p. 201, Editions Masson, 1992.

[CAG] G. CAGNAC, E. RAMIS, J. COMMEAU, *Nouveau cours de Mathématiques Spéciales, Application de l'Analyse à la Géométrie*, p. 272-278, Editions Masson, 1963.

[CAR] G.F. CAREY, W.B. RICHARDSON, C.S. REED, B.J. MULVANEY, *Circuit, Device and Process Simulation, Mathematical and Numerical Aspects*, p. 155, John Wiley & Sons, 1996.

[CHA] M.V.K. CHARI, P.P. SILVESTER, *Finite Elements in Electrical and Magnetic Field Problem*, John Wiley & Sons, 1980.

[COU] J.L. COULOMB, J.C. SABONNADIERE, *CAO en Electronique*, Editions Hermès, 1985.

[DAU1] R. DAUTRAY, J.L. LIONS, *Analyse mathématique et calcul numérique pour les sciences et techniques*, vol. V, p. 368, Editions Masson, 1988.

[DAU2] R. DAUTRAY, J.L. LIONS, *Analyse mathématique et calcul numérique pour les sciences et techniques*, vol. V, p. 451, Editions Masson, 1988.

[DAU3] R. DAUTRAY, J.L. LIONS, *Analyse mathématique et calcul numérique pour les sciences et techniques*, vol. V, p. 479, Editions Masson, 1988.

[DAU4] R. DAUTRAY, J.L. LIONS, *Analyse mathématique et calcul numérique pour les sciences et techniques"*, vol. V, p. 933-943, Editions Masson, 1988.

[DUR] E. DURAND, *Electrostatique*, vol. III, p. 115-146, Editions Masson, 1966.

[FOU] G. FOURNET, *Electromagnétisme, lois générales*, p. 7-12.

[HAR] R.F. HARRINGTON, *Time Harmonic Electromagnetic Fields*, McGraw Hill, 1961.

[JAC1] J.D. JACKSON, *Classical Electrodynamics*, Chapter V, p. 169-173, Chapter VI, p. 210-213, John Wiley & Sons, New York, 1975.

[JAC2] J.D. JACKSON, *Classical Electrodynamics*, Chapter I, p. 29-30, John Wiley & Sons, New York, 1975.

[JAC3] J.D. JACKSON, *Classical Electrodynamics*, Chapter VI, p. 245-248, John Wiley & Sons, New York, 1975.

[JAC4] J.D. JACKSON, *Classical Electrodynamics*, Chapter VI, p. 249-252, John Wiley & Sons, New York, 1975.

[JAC5] J.D. JACKSON, *Classical Electrodynamics*, Chapter II, p. 66, John Wiley & Sons, New York, 1975.

[JEN] A. JENNINGS, *Matrix Computation for Engineers and Scientists*, John Wiley & Sons, 1977.

[LAN1] L. LANDAU, E. LIFCHITZ, *Electrodynamique des milieux continus*, Chapter IX, p. 325-332, Editions Mir, Moscow, 1969.

[LAN2] L. LANDAU, E.LIFCHITZ, *Mécanique quantique (théorie non relativiste)*, Chapter III, problem 4, p. 93-94, Editions Mir, Moscow, 1967.

[LAN3] L. LANDAU, E. LIFCHITZ, *Mécanique des fluides*, p. 242-248, Editions Mir, Moscow, 1971.

[LAN4] L. LANDAU, E. LIFCHITZ, *Electrodynamique des milieux continus*, p. 69, Editions Mir, Moscow, 1969.

[LAS1] P. LASCAUX, R. THEODOR, *Analyse numérique matricielle appliquée à l'art de l'ingénieur*, vol. I, Chapter 1, p. 62, Editions Masson, 1986.

[LAS2] P. LASCAUX, R. THEODOR, *Analyse numérique matricielle appliquée à l'art de l'ingénieur*, vol. I, Chapter IV, p. 207-243, Editions Masson, 1986.

[LAS3] P. LASCAUX, R. THEODOR, *Analyse numérique matricielle appliquée à l'art de l'ingénieur"*, vol. I, Chapter I, p. 52-53, Editions Masson, 1986.

[LEF] S. LEFEUVRE, *Hyperfréquence*, Chapter I, p. 14, Editions Dunod.

[MER] B. MERCIER, *An Introduction to the Numerical Analysis of Spectral Methods*, p. 43-55, Springer-Verlag, 1988.

[MEY] Y. MEYER, *Ondelettes et opérateurs*, vols. I and II, Editions Hermann, 1990.

[MOS] J.R. MOSIG, *Numerical Techniques for Microwave and Millimeter Wave Positive Structures*, John Wiley & Sons, New York, 1989.

[RAV1] P.A. RAVIART, J.M. THOMAS, *Introduction à l'analyse numérique des équations aux dérivées partielles*, Chapter VI, p. 144-146, Editions Masson, 1988.

[RAV2] P.A. RAVIART, J.M. THOMAS, *Introduction à l'analyse numérique des équations aux dérivées partielles*, Chapter IV, p. 79-95, Editions Masson, 1988.

[ROC] Y. ROCARD, *Electricité*, Chapter 1, p. 159-164, Chapter 3, p. 182-186, Editions Masson, 1956.

[ROU] E. ROUBINE, *Compléments d'Electromagnétisme*, Ecole Supérieure d'Electricité, 1980

[SCH] L. SCHWARTZ, *Méthodes mathématiques pour les sciences physiques*, Chapter II, p. 76-90, Editions Hermann, 1965.

[SCHE] M.R. SCHEINFEIN, J.C. LIAO, O.A. PALUSINSKI, J.L. PRINCE, "Electrical performance of high-speed interconnect systems", *IEEE Transactions on Components, Hybrids, and Manufacturing Technology*, vol. chmt-10, 303-309, 1987.

[VAL1] G. VALIRON, *Théorie des fonctions*, Chapter IX, p. 252, Editions Masson, 1948.

[VAL2] G. VALIRON, *Théorie des fonctions*, Chapter IX, p. 227, Editions Masson, 1948.

[VAP1] A. VAPAILLE, *Physique des semi-conducteurs*, vol. I, Chapter 2, p. 35-43, Editions Masson, 1970.

[VAP2] A. VAPAILLE, *Physique des semi-conducteurs*, vol. I, p. 140, Editions Masson, 1970.

[VEN] J. VENKATARAMAN, S.M. RAO, A.R. DJORDJEVIC, T.K. SARKAR, Y. NAIHENG, "Analysis of arbitrarily oriented microstrip transmission lines in arbitrarily shaped dielectric media over a finite ground plane", *IEEE Transactions on Microwave Theory and Techniques*, vol. mtt-33, 952-960, 1985.

[WAI] R. WAIT, A.R. MITCHELL, *Finite Element Analysis and Applications*, John Wiley & Sons, Chichester, 1985.

[WAR] A. WARUSFEL, *Techniques de l'Ingénieur*, volume A1, article A101-1 (normes et distances).

[WEI] C. WEI, R.F. HARRINGTON, J.R. MAUTZ, T.K. SARKAR, "Multiconductor transmission lines in multilayered dielectric media", *IEEE Transactions on Microwave Theory and Techniques*, vol. 32, 439-450, 1984.

Index

3D wavelets, 256, 261

A, B

Ampere's theorem, 23, 24
Axial
　pseudo-vector distributions, 11
　vector distributions, 1, 9
Backward waves, 131, 132, 133, 134, 136
Banach
　algebra, 220
　fixed point theorem, 224, 227
　space, 220
　vector space, 222
Bent
　media, 127, 129
　stratified media, 52
　transmission lines, 86, 127
Bessel
　functions, 209, 228
　modified functions, 208
Biaxial coordinates, 272, 274
Bifurcations, 131, 151, 156, 216
Boundary conditions, 25, 27, 28, 30, 31, 41

C

Capacitance matrix, 116
Characteristic impedance, 137, 138, 139, 143, 151, 161, 189, 190, 196, 198
Charge surface density, 27

Complex networks, 137, 167, 216
Compounded matrix functions, 179, 180, 182, 186
Convolution method kernel, 208, 209
Conductance matrix, 117
Continuity equations, 21
Convolution
　kernel, 4
　product, 4, 8, 9
Crosstalk, 131, 146, 147, 157
　impedance matrices, 194
Current
　bulk density, 26
　displacement bulk density, 26
　surface density, 26, 27, 42, 43
Curvature vector, 59
Cut-off wavelength, 91, 92, 105, 107, 109
Cylindrical wavelets, 277

D

Differential operator, 13, 16
Digital
　signal expansion, 235
　signal processing, 214
　signals propagation, 226
Dirac
　distribution, 5, 6, 8, 9, 12
　pulse, 189, 207, 210
　pulse distortion, 207
Discretization method, 32

Duality, 25, 32, 34, 36, 40
   between electrical and magnetic currents, 64
   in electromagnetism, 15, 25, 38, 40, 69, 79, 82
   relations, 65
Duncan-Sylvester
   approach, 196
   formula, 160, 192, 197
   spectral expansion, 134, 135

E

Electrical
   current density, 10, 11, 15, 22
   field, 11, 15, 19, 24
   induction, 11, 21, 25
   scalar potential, 36
   stacked transmission line behavior, 67
   vector potential, 35, 38
Electromagnetic
   images, 25, 29
   influence matrix, 143
   interference, 25, 44, 45, 46, 47, 48, 51, 52
   interference, tree of, 48
Elementary solution, 8, 9
Elliptic coordinates, 274
Equipotential line, 116

F

Faraday-Lenz theorem, 24
Finite difference, 163, 170
   method, 170, 174, 178
Forward wave, 132, 133, 136
   and backward waves, 131
Foucault currents, 196
Fourier transform, 89, 90, 92, 93, 94, 97, 123, 189, 211, 212, 213, 214, 217, 218
Frenet trihedron, 51, 85

G, H

Gauss theorem, 25
   in electrostatics, 96, 110
Green's
   formula, 25, 31
   kernels, 2, 9, 40, 41, 118, 120
Hamiltonian operator, 70, 72, 84
Heterogenous
   coupling, 150, 151
   media, 38, 40, 41, 52, 60, 68, 71
   stratified media, 163, 183
Hilbertian
   norm, 3, 91
   scalar product, 30
Hölderian property, 172
Homogenous
   coupling, 143, 161
   media, 85, 86, 90, 105, 113, 114, 117, 128

I, K

Impedance
   matching, 146, 189, 199
   matrix, 199
Incident
   and reflected waves, 137, 138
   field, 29
Inductance matrix, 117
Influence matrix, 143
Initial conditions, 25, 27, 28, 32
Integral equations, 25, 32, 33, 43
Interconnection
   between points called "nodes, 44
   network, 34, 44, 52
Isotropic mapping, 180
Isotropy, 4, 17, 19
Kirchoff's laws, 131, 151, 153

L

Laplace transform, 190, 191, 203, 205, 206, 207, 212

Laplacian operator, 108
Linear operator, 119, 125
Lineic
  capacitors, 269
  parameters, 118
Lorentz
  condition, 102, 111
  gauge, 36, 37, 38
Lossy
  coupled lines, 243, 246, 250, 254
  propagation equation, 36, 67, 68, 89, 94
  transmission line, 125
  transmission line equation, 126
  wave propagation, 85
  wave propagation matrix operator, 244

M

Magnetic
  current density, 11, 15
  field, 1, 10, 11, 15, 16, 19, 23, 24
  induction, 22, 24
  pseudo-scalar, 36
Matrix
  differential operator, 98
  drift equation, 183
  operators in the wavelet space, 251
  transmission line equation, 131, 135
  velocity operator, 179, 183, 187
Maxwell's equations, 1, 2, 4, 5, 15, 19, 20, 21, 22, 23, 25, 26, 28, 32, 34, 44
Method moments, 30
Microstrip, 88
Microstrip, 88, 90
Modal
  analysis, 85
  characteristic impedance matrix, 196
  crosstalk impedance matrix, 196
Modified relaxation method, 275
Moving wave, 138, 139

Multiple reflections, 139, 141, 146, 250
  recursive equation, 174
Multiplicity order, 141, 143

N, O

Nabla operator, 14
Newton's
  centered scheme, 172
  down scheme, 171
  up scheme, 172
Nodal
  matrices, 164, 174
  operational matrices, 163, 167, 177
Nodes, 44
Optical fiber, 283, 284
Oriented graphs, 25, 44, 45

P

Parabolic
  antenna, 284
  coordinates, 271
Parseval theorem, 214
Physical data, 263
Picard method, 224, 227
Poisson equation, 22, 27, 113, 116, 118, 119
Polar vector distribution, 10, 11, 23
Potentials in electromagnetism, 34
Projection operator, 171, 174
Propagator, 131, 136, 137, 138, 144, 145, 162
  group, 137
Pseudo
  tensor distributions, 9
  vector distribution, 9, 22

Q, R

Quadripole, 174
Quasi-TEM, 86, 118, 119, 125, 127, 129
Radius
  of curvature, 59
  of twisting, 63

Recursive-differential equation, 168
Reflected
    fields, 28, 29, 30
    waveforms, 162
Riesz-Frechet representation theorem, 3

S

Sampling, 212, 214, 218
Scalar and vector potentials, 34
Scattering matrix, 146
Schwarz theorem, 8, 13, 16, 20
Shannon interpolating formula, 189, 213
Skin depth, 209, 210
Space of wavelets, 243
Spectral
    analysis, 51, 71, 85
    expansion, 137, 160, 161, 192
    expansion of matrix functions, 192
    theory of matrices, 189
Spectrum range, 189, 214, 215, 216, 217, 218
Standing waves, 137
State
    equation, 131, 156, 159, 162
    vectors, 156
Stoke formula, 23, 24
Stratified
    media, 51, 52, 53, 67, 70, 80, 85
    heterogenous media, 182
Stripline, 95
Symmetry, 3, 4, 5, 9, 11, 15, 17, 18

T

Tchebychev functions, 236, 237
Technological data, 267
TE mode, 66, 82, 83, 84
Telegrapher
    equation, 85, 91, 92, 98, 99, 109, 111, 112, 230, 231, 233
    operator, 98

TEM
    approximation, 243
    mode, 66, 92, 104, 105, 109, 111, 112
Tensor distribution, 9
Test functions, 32, 33, 43
Theory of distributions, 1, 5, 6, 10, 15, 19, 25, 26, 28
Time step, 174
TM mode, 82, 83, 84
Topological
    cylinder, 47, 48
    modeling, 44, 47
    net, 47, 51
    surgery, 51
Topology of electromagnetic interferences, 43, 46
Transfer matrix, 189, 191, 194, 195, 196, 197, 198, 199, 213
Transient state time, 159, 160, 161, 168, 170
Transitive enclosure, 46
Transmission line
    equations, 51, 53, 76, 116, 164
    matrix equation, 80, 85, 167, 188
    theory, 85
Truncature criterion, 216
Twisting vector, 62

V, W

Variational method, 70, 71, 107
Vector
    analysis, 1, 12
    distribution, 9, 13, 14, 15, 19, 21, 22
    potentials, 25, 34
Von Neumann algebra, 224
Waveguide, 85, 91, 103, 104, 105, 283, 284
Wavelet
    method, 219, 243, 244, 255
    seed, 223, 224, 231, 260